Smart Materials Taxonomy

Smart Materials Taxonomy

Victor Goldade

Francisk Skorina
Gomel State University
Gomel, Belarus;
V.A. Belyi Metal-Polymer
Research Institute of National
Academy of Science of Belarus
Gomel, Belarus

Serge Shil'ko

V.A. Belyi Metal-Polymer
Research Institute of National
Academy of Science of Belarus
Gomel, Belarus

Alexander Neverov

Belorussian State
Transport University
Gomel, Belarus

CRC Press
Taylor & Francis Group
Boca Raton London New York

CRC Press is an imprint of the
Taylor & Francis Group, an **informa** business

A SCIENCE PUBLISHERS BOOK

CRC Press
Taylor & Francis Group
6000 Broken Sound Parkway NW, Suite 300
Boca Raton, FL 33487-2742

First issued in paperback 2020

ISBN-13: 978-1-4987-1619-2 (hbk)
ISBN-13: 978-0-367-73776-4 (pbk)

Library of Congress Cataloging-in-Publication Data

Goldade, V. A. (Viktor Antonovich)
    Smart materials taxonomy / Victor Goldade, Serge Shilko, Alexander Neverov.
      pages cm
"A CRC title."
Includes bibliographical references and index.
ISBN 978-1-4987-1619-2 (hardcover : alk. paper) 1. Smart materials-- Classification.
2. Nanostructured materials-- Classification. I. Shil'ko, Serge. II. Neverov, A. S. (Aleksandr Sergeevich) III. Title.

TA418.9.S62]G65 2015
620.1'12-- dc
          23 2015028949

Visit the Taylor & Francis Web site at
http://www.taylorandfrancis.com

and the CRC Press Web site at
http://www.crcpress.com

Dedication

We dedicate this book to our untimely deceased friend and colleague Prof. Leonid Pinchuk. Chapter 4 was written by him.

Authors

Preface

The terms 'smart materials' and 'intelligent materials', as well as the occasionly used term 'clever materials', were introduced to describe the special functionality of materials in performing interaction between components of electronic devices and electromagnetic fields. The problem of designing miniature electronic systems integrated directly into the material structure seemed almost insoluble until recently. Accessible techniques of solving this have been developed due to achievements of microelectronics and nanotechnology dealing with nanodimensional structural components. Therefore, two key elements of modern materials science are 'nano' and 'smart' materials.

The range of smart materials has been naturally expanding. At present the response of materials to external influence has been implemented in the entire combinations of phenomena existing in nature. Scientific publications and the internet have accumulated an enormous amount of uncombined, often promotional information about smart materials. The challenge emerged to systematize (and at the first opportunity – to formalize) this information from getting outdated and continuously replenished.

The aim of the book is to systematize the complete accessible (and even potential) integrity of smart materials and engineering systems with the techniques of taxonomy based on the attributes inherent to cybernetic systems. We have done our best to avoid the touch of sensation accompanying a considerable number of publications containing the information about smart materials, and provide the scientific justification of the techniques of their systematization and description. Significant attention has been paid to modeling expedient reactions of smart materials to external influence by methods of mechanics, as well as to the prediction new specific effects and relevant heterogeneous structures with feedbacks.

We express our heartfelt gratitude to Prof. Yu.M. Pleskachevsky – for his valuable observations, which were taken into account in the previous version of the manuscript, to Prof. V.N. Kestelman for assistance in publishing this book, as well as to our colleagues, researchers of the Metal-Polymer Research Institute of the National Academy of Sciences of Belarus, S.V. Zotov, E.M. Petrokovets and T.V. Ryabchenko – for their invaluable service in preparing this treatise.

Victor Goldade
Serge Shil'ko
Alexander Neverov

Gomel
March 2015

Contents

Introduction

The high standards of development in modern society are achieved largely due to the implementation in high-tech products of the ideas and methods of cybernetics, which is the science of general laws of obtaining, storing and processing of information. A cybernetic system, when considered independently of its material nature, consists of interrelated components that collect, store, process and exchange information. Such systems are integrated into the structure of all major technological objects, including materials, which have been assigned the function of the cybernetic system since the second half of the 20th century.

The terms 'smart materials' and 'intelligent materials' were proposed due to the need of special materials, which allowed interaction between the components of electronic devices used to transfer, process and store information by means of electromagnetic fields. Since the early 50s of the last century, solid state electronics has been progressing considerably based on the semiconductors. Microelectronics emerged as the most promising trend in the 1960s, followed by molecular electronics in the 1990s. Molecular electronics operates with molecules functioning as electronic device components. The challenge of designing miniature electronic systems integrated into the material structure has seemed almost impossible until recently. At present, due to the advances in microelectronics and nanotechnology, techniques are available to manipulate structural components as small as 0.1–100 nm. Therefore, the two key terms of the materials science in the 21st century are 'nano' materials and 'smart' materials.

That resulted in the interdisciplinary infrastructure formation of the materials science of the 'smart' materials, and nowadays specialists omit quotation marks for the word smart. The exact definition of the term "smart materials" will be provided later, meanwhile let us note that a smart material performs its function in the technological article by exhibiting the attributes of artificial intelligence. It recognizes changes in its own structure when in operation; it processes this information and automatically corrects the structure by initiating secondary physical and chemical processes by using the feedback system through the energy provided by the external influence. The feedback system is among the basic notions of cybernetics, which stipulated the adaptability of natural objects and technology to external conditions. The feedback produces the signals of physical, chemical or biological nature, which makes the material structure to adapt the external conditions in order to extend the article's life in operation.

The range of smart materials created during the growth of the electronic technology naturally kept expanding. At present, smart materials can respond to any combinations of phenomena existing in nature. The groups of smart materials, such as tribological, anticorrosive, sealants, etc. for machine elements has progressed vigorously. Numerous smart systems have appeared in medicine and biotechnology. So far, the scientific publications and the internet have accumulated an enormous amount of scattered, frequently promotional information about smart materials. The challenge emerged to systematize this information, some of which becomes outdated while it is being continuously replenished. The urgency to create a system persists, even though over a hundred books including numerous monographs on design and application of smart materials has been published (Gandhi and Thompson 1992, Wang 1998, Tomlinson and Bullough 1998, Lazzari et al 2002, Watanabe and Ziegler 2003, Schulz et al 2005, Singh 2005, Ritter 2006, Varadan et al 2006, Richtering 2006, Reece 2007, Van Langenhove 2007, Leo 2007, Schwartz 2008, Luk'yanchuk and Mezzane 2008, Zhao and Ikeda 2009, Varadan 2009, Laso 2009, Ananthasuresh et al 2012, Agrawal et al 2013). These references are given here to demonstrate the variety of problems accompanying the development of smart technological systems. Most of these publication are referred to later in the text wherever necessary.

Even though there is an ever increasing interest in this topic, so far there has been no universal classification of smart materials, which is practical for engineers and technologists to use. The structural and functional diagram of smart systems borrowed from the materials science as the triad of 'sensor-actuator-processor' (Gandhi and Thompson 1992) needs material filling. The smart material models remain incomplete without the account of physical and chemical processes which are, firstly, its natural response to external influence and, secondly, initiated by the feedback consuming the energy of external influence. The multitiude of the factors, to be taken into account when systematizing smart materials, dictated the domination of the descriptive approach to solving the problem. An example of systematizing endeavor is represented in (Pleskachevskii 2008), where the smart materials are systematized based on the criteria of structural attributes and links, functional characteristics, principles of self-organization and numerous other criteria which nevertheless, ignore the nature of physical and chemical processes evolving in materials, energy and technological aspects of their formation.

The objective of the present publication is to systematize the body of knowledge on the smart materials and technological systems with the techniques of taxonomy based on the attributes inherent to cybernetic systems. This approach enables us to describe the whole constellation of the smart materials which are already available and those that in the development of a single table, justify the three-stage development process: registration of changes in the material structure in operation, the choice of mechanism of correcting the structure using the feedback, implementation of the chosen mechanism by applying the achievements of natural (physical, chemical, biological) sciences.

The authors did their best to avoid the touch of sensation accompanying a considerable number of publications containing the information about smart materials and provide the scientific justification of the techniques of their systematization and modeling. It is the authors' belief that this publication will allow one to promote the development of smart materials from the category of heuristic inspiration to the professional engineering and technological level, permitting the specialists to develop smart materials and technological systems in the process of daily routine work.

References

Agrawal, C.M., J.L. Ong and R.M. Appleford. 2013. *Introduction to Biomaterials: Basic Theory with Engineering Applications*. Cambridge University Press, Cambridge.

Ananthasuresh, G.K., K.J. Vinoy, S. Gopalakrishnan, K.N. Bhat and V.K. Aatre. 2012. *Micro and Smart Systems: Technology and Modeling*. John Wiley & Sons, New York.

Gandhi, M.V. and B.S. Thompson. 1992. *Smart Materials and Structures*. Springer, Berlin.

Laso, M. 2009. *Smart Materials Modelling*. John Wiley & Sons, New York.

Lazzari, B., M. Fabrizio and A. Morro. 2002. *Mathematical Models and Methods for Smart Materials*. World Scientific Publishing Co., New Jersey.

Leo, D.J. 2007. *Engineering Analysis of Smart Material Systems*. John Wiley & Sons, New York.

Luk'yanchuk, I.A. and D. Mezzane. 2008. *Smart Materials for Energy, Communications and Security*. Springer, Berlin.

Pleskachevskii, Yu.M. 2008. Main tendencies of new ideas evolution in science of materials. Materials, Technologies, Tools 13(3): 5–13.

Reece, P.L. 2007. *Progress in Smart Materials and Structures*. Nova Science Publishers, New York.

Richtering, W. 2006. *Smart Colloidal Materials (progress in colloid and polymer science)*. Springer, Berlin.

Ritter, A. 2006. Smart Materials: In Architecture, Interior Architecture and Design. Springer, Berlin.

Schulz, M.J., A.D. Keklar and M.J. Sundaresan. 2005. *Nanoengineering of Structural, Functional and Smart Materials*. Taylor & Francis, London.

Schwartz, M.M. 2008. *Smart Materials*. CRC Press, Boca Raton.

Singh, J. 2005. *Smart Electronic Materials: Fundamentals and Applications*. Cambridge University Press Cambridge.

Tomlinson, G.R. and W.A. Bullough. 1998. *Smart Materials and Structures.* Taylor & Francis, London.

Van Langenhove, L. 2007. *Smart Textiles for Medicine and Healthcare: Materials, Systems and Applications.* CRC Press, Boca Raton, Florida.

Varadan, V.K. 2009. *Handbook of Smart Systems and Materials.* Taylor & Francis, London.

Varadan, V.K., K.J. Vinoy and S. Gopalakrishnan. 2006. *Smart Material Systems and MEMs: Design and Development Methodologies.* John Wiley & Sons, New York.

Wang, Z.L. 1998. *Functional and Smart Materials: Structural Evolution and Structure Analysis.* Kluwer Academic Publishers, Boston.

Watanabe, K. and F. Ziegler. 2003. *Dynamics of Advanced Materials and Smart Structures.* Springer, Berlin.

Zhao, Y. and T. Ikeda. 2009. *Smart Light-responsive Materials: Azobenzene-containing Polymers and Liquid Crystals.* Wiley-Interscience, New York.

List of abbreviations

ABS – antilock brake system
AMR – active magnetic regeneration
BAP – bioactive point
CI – corrosion inhibitor
EME – electromagnetic emission
EMF – electromagnetic field
EMR – electromagnetic radiation
EMS – electromagnetic screen
FBS – feedback system
FE – filtering element
FM – frictional material, filtering material
GMR – giant magnetoresistance
HOA – hydroxyapatite
HTSC – high-temperature superconductor
JI – joint implant
MA – medicinal agent (drug)
MCE – magnetocaloric effect
MIC – microelectronic integrated circuit
MM – metamaterial
MO – magneto-optical
MVF – microwave frequency
PhC – photo-conductivity
PT – physiotherapy
PTFE – polytetrafluoroethylene
RAM – radioabsorbing material
RFID – radio frequency identification
SEM – scanning electron microscopy
SL – synovial liquid
SM – smart material
TCR – temperature coefficient of resistance
UHF – ultrahigh-frequency
UHMPE – ultrahigh molecular polyethylene
T_b – boiling temperature
T_m – melting temperature

chapter one

Concept of smart materials

The smart material (SM) differs from common material the components of its structure function as the cybernetic system efficiently adapting the material to the varying operation conditions. Humanity has gone a long way from using the materials prepared by nature to creating the SM. The notions of materials science are formulated below preceding the SM definition, supplementing this term as well as the cybernetic notions used to describe the SM. The energy characteristics of external influence are given which the technological materials experience in operation. The energy of this influence is used by the SM to 'repair' the operation damage.

1.1   Retrospective review

The history of evolution of materials is dialectically linked with the social history. The drive to discover new materials in all historic epochs was due to the human aspiration to improve the quality of lye life.

The historic stages are named after the materials, which 'created the epoch' (stone, copper, bronze and iron ages of civilization) reflecting what they meant in the history of humanity since ancient times. The next generation of materials began its life cycle in the second millennium a.d. when the falling water energy was used to drive machinery. Rapid metallurgy progress changed the engineering radically together with the technology of materials. This epoch ended in the 20th century by developing three main processes of steel production of named after their inventors Bessemer, Marten and Thomas. On the boundary between 19th and 20th centuries, plastics were produced first and played an outstanding role in the technological history. Nevertheless, the expected 'age of plastics' never came across. The appeal of ceramics first used in the Neolithic age was revived after the 1930s when powder metallurgy was developed successfully. New types of materials which the humankind continues to use acquired public recognition. Scientific and technological revolution in the late 1940s led to creation of semiconductors, superconductors, synthetic diamonds and other fundamentally novel materials by developing new raw material facilities.

At this time the scientists of materials had formed the final opinion that natural resources of availability of traditional technical materials have been exhausted. The first to experience that was the designers of machinery operating in extreme conditions: elevated temperatures (engines), deep vacuum (outer space missions), super fast friction (weapons) and others. The main trend of the science of materials included composite materials by combining components of various natures in the second half of the 20th century. The present tendency of improving composite materials is to produce them by nanotechnology. Now the SM are the best engineering materials, however they are still few.

The advent of SM is assumed to be in 1834 when the French physicist and mathematician A. Ampere, one of the founders of electrodynamics, wrote a book *Essays about the philosophy of sciences* (*Essais sur la Philosophic des Sciences*). It listed all sciences known at that time. The science named 'Cybernetics' was listed number 83. Ampere called this science of techniques of social management cybernetics using the Greek term of the ancient time applied to the navigation art ('kybernětes' – helmsman).

Cybernetics determined now as the science of general laws of acquiring, storing, transmitting and processing information (Glushkov 1973) occurred in the USA when the computer was developed. Norbert Wiener, professor of mathematics at the Massachusets Technological University, used the word 'cybernetics' in his book authored in 1948 (Wiener 1948). It summarizes the regularities relating to the biological, engineering and social systems of management (the latter were discussed particularly in the book *Cybernetics and Society*, published in 1954). This wonderful event in the history of science was prepared by numerous specialists in the spheres of mathematics, mechanics, automatic controlling, computing, and physiology of higher nervous activity.

The first ideas of regulation as system subordination to definite order have originated in biology. The hypothesis of reflex responses had contained the feedback principle premise. René Descartes, French philosopher, mathematician, physicist and physiologist, introduced (1664) the notion of reflex. Jirzhi Procházka, Czech anatomist and physiologist, expanded the notion believing the nervous system was a moderator between organism and environment (1784). Charles Bell, Scotch anatomist, surgeon and physiologist, concluded about the reflex principle of nervous system functioning (1814).

Thus, the base of modern ideas of systems of regulation in living organisms was founded. The famous Russian physiologist I.M. Sechenov developed the principles of the theory of reflexes (1866) and advanced the swneering idea 'the conception of mechanistic brain functioning is a treasure for physiologist'. Nobel Prize winner (1904) I.P. Pavlov developed the theory of high nervous activity. Through his experiments he established the main characteristics of conditioned reflexes and proposed this term. Pavlov believed that man was a system subordinated to universal natural laws but unique in consequence of highest self-regulation degree because he maintained, restored, corrected and perfected himself.

N.A. Bernstein, Soviet neurologist and psychophysiologist, created (1941) the theoretical principles of modern biomechanics anticipated a number of cybernetics regulations. The theory of functional system advanced by Russian physiologist P.K. Anokhin (1965) contains the idea of feedback of nervous systems. Ukrainian biologist I.I. Shmalhausen (1967) applied the theory of feedback to the population genetics. The works of I.P. Pavlov and his followers laid the base of biological cybernetics which British biologist W.R. Ashby represented as a science (1959) in the monograph *Introduction to Cybernetics.*

I.A. Vyshnegradskii, Russian mathematician and mechanical engineer, laid the foundations of the theory of automatic regulation and the theory of stability of regulation systems (1878). A.M. Lyapunov, one of the greatest mathematicians of the 19th century, was the first to solve (1892) general problems of motion stability, which are the foundation of the modern theory of automatic management.

The material base of systems control implementing the cybernetic techniques was the electronic computing engineering. The 'analytic machine' of British mathematician Charles Babbage who anticipated the idea of computer (1835) is considered a prototype of modern digital computer. Another British mathematician A.M. Turing created the theory of automatic devices in 1936. His automaton suitable in principle to implement any algorithm, is known as the 'Turing machine'.

The cybernetics in the Soviet Union was branded as the false science and capitalist housemaid. It was rejected for a decade before Wieners book was translated into Russian in 1958. Regretably, the 'father' of the cybernetics who revolutionized social life, was not awarded the Nobel Prize.

N. Wiener and creator of digital computer American mathematician J. von Neiman combined the ideas of I.P. Pavlov with the engineering principles implementing artificial intelligence. Later it was established that the structure of nerve fibers of man consisting of nerve cells (neurons) is similar to the engineering control systems in the structure of links. Both these systems accumulate and process information discretely by the principle 'yes-no'. George Bool, one of the founders of mathematical logics, applied the binary calculation to the mathematical and logic apparatus and developed in 1854 the Boolean algebra. It is used to estimate information and construct the smart systems.

SM are cybernetic systems possessing numerous internal links subjected to stochastic (random) variations. If the external influence on the SM yields N different results, out of which n are favorable, the probability of a favorable result is expressed by the fraction $P = n/N$. American engineer and mathematician C.A. Shannon, one of the authors of information theory, proposed in 1948 to estimate the unequally probable results with the equation of information entropy:

$$i = -\left(P_1 \lg_2 P_1 + P_2 \lg_2 P_2 + ... + P_n \lg_2 P_n\right) = -\sum_{j=1}^{n} P_j \lg_2 P_j,$$

This equation is similar to the entropy formula of Boltzman (the Austrian physicist, one of the founders of statistical physics in 1872). Shannon introduced the information entity – the *bit* (from the *binary digit*) assuming the values of 0 and 1.

When retrospecting the SM, it is worthwhile to mention A.A. Bogdanov who published in 1913 the monograph *Universal organizational Science* (*Tectology*) in which he advanced original ideas much ahead of the stipulations of modern cybernetics.

When creating the SM theory, a noticeable role was played by the concept of 'black box' introduced by N. Wiener; it implies the system with unknown internal device and the processes evolving in it. The methodology of the 'black box' exploration assumes the study of reactions to the assigned input signals. Wiener used the 'black box' principle to select the working hypothesis about the assumed solar system structure by comparing its behavior and that of the known model. This method of 'black box' illumination came to be used after that to model the SM structure.

The evolution of the SM theory is linked with the research of many Russian scientists: admiral-engineer A.I. Berg, the president of the cybernetic scientific board of the USSR Academy of science in the 1970s; V.M. Glushkov, the author of the theory of digital automatons; L.V. Pontryagin who developed the principle of optimality of cybernetic systems (Pontryagin's principle of maximum) and A.N. Kolmogorov, founder of scientific schools of the theory of probabilities and the theory of functions.

The SM development acquired interdisciplinary nature in the 20th century. The research of industrial SM application at almost all laboratories of Xerox Co. made it leader in this trend. SM development is included in the list of promising projects by Hewlett-Packard Co. US Army Research Lab. and special subdivisions of the University of Delaware conduct SM development for army weapons.

Miniaturization of electronic units has reached the limit beyond which the evolving spin effects should be taken into account. It ushered a promising new sphere of SM applications as materials of spin electronics or spintronics using the spin of quantum particles as information carriers (Marrows et al 2009). The discovery of the first considerable spintronic effect – the giant magnetic resistance –was awarded the Nobel prize in 2007 (German physicists A. Fert and P. Grunberg).

The results of SM research has been an actual subject in many scientific journals. 'Smart Materials Bulletin' is being published for 20 years. 'The Intelligent materials' is a popular journal. There are permanent rubriques dedicated to the SM, among them 'Advanced Composite Bulletin',

'Advanced Ceramics Report', 'Applied Surface Science', 'Biomedical Materials', 'Materials Science and Engineering: B, C', 'Journal of Alloys and Compounds', 'Materials and Wave Interaction', 'Journal of Magnetism and Magnetic Materials', 'Polymer Cells and Networks', 'Progress in Materials Science', 'Progress in Polymer Science', 'Computational Materials Science', 'Microelectronic Engineering', 'Nanostructured Materials', 'Photonics and Nanostructures: Fundamentals and Applications' (Elsevier), 'European Physical Journal' (Springer), 'Philosophical Magazine' (Taylor & Francis), 'Nano' (World Scientific) and 'Journal of Advanced Materials' (Nova Science Publishers Inc.). It proves that the development of SM and smart engineering systems has become the leading trend of materials science in the 20th century.

1.2 Main notions

Matter is the substance type, the integrity of discrete formations possessing the mass at rest. Man uses the matter as a base of production means (in other words, the work means and objects) to optimizing material benefit. The engineering uses the term 'material' instead of the notion 'substance base'. Materials make up both the means of work and objects. The means of work (machinery, equipment, tools, buildings, communication means, cargo transportation means, and land as universal work means) serve to influence the work objects – all that is what the man work targets.

The materials in the engineering articles receive and transmit mechanical loading, conduct electricity and heat, create a barrier to liquid and gases, etc. On point of complication the operating functions, the categories of materials form a sequence: simple – multifunctional – adaptive – active – smart (intelligent).

Simple materials perform only one main operating function in the engineering article. The word 'main' implies that almost always the entire set of physical, chemical, mechanical and other natural properties of the material are proceeded uncontrolled in the article infusing it with some additional operating properties.

Multifunctional materials perform in the main article and have additional functions. For instance, bronze is structural and triboengineering material in machine building; some bronze grades feature stronger corrosion resistance. Many constructional plastics serve to protect metallic pieces from corrosion and are used for electric insulation. Most ceramic materials combine exceptional hardness, dielectric properties, as well as thermal and chemical endurance, explaining multifunctional application of ceramics in engineering.

Adaptive (accommodative) **materials** change their structure expediently under the effect of operating factors (mechanical stress, temperature, physical fields, environment) when their intensity reaches a threshold.

In principle, the structure of any material changes naturally under the external influence. 'Expediently' implies that the structure change of adaptive material promotes the applicability of the article which contains it. In other words, new properties of the material 'adapted' to changed operational conditions, determine the higher competitiveness and quality of the article. The adaptive material structure can restore its initial state when the external influence ceases (Shil'ko 2011); otherwise, the change on structure becomes irreversible.

The example of the adaptive material is plastic grease, which is a lubricant like material obtained by the introduction of the solid thickener (soap, paraffin, carbon black and others) into liquid petroleum or synthetic oil. The lubricant acquires solid body properties at loads below the strength limit of the dimensional framework produced by the thickener. Under heavy load, it turns into an abnormally viscous lubricating fluid. After the load ceases, the framework structure is restored and the lubricating material becomes like a solid body again.

Active materials perform their inherent engineering functions and additionally impact positively of physical, chemical or biological nature influence to the interfaced parts of the article and the environment. The criterion of usefulness is the promoted applicability, quality and competitiveness of the article.

For example, chemically active filtering material comprises the carbon fabric impregnated with complexions – amino-polycarbonic acids and their derivatives. The complexions are remarkable because they enter into reactions with cations of metals producing stable coordination compounds. The filtering material demonstrates activity by chemically combining the ions of heavy metals solvated in the filtering fluid. The heavy metals are hazardous for environment and contain impurity hard to perceive.

Smart materials (SM) perform the operating functions relevant to its natural properties in the engineering system, but once the external energy reaches some threshold, it transforms this energy into its structural changes, improving the level of operating properties in the first place and secondly it controls and automatically adjusts this level through the feedback system by comparing the value of external influence and the extent of changing properties.

Thus, the distinctive SM attributes are the following:

1. the SM operating functions, like those of any material, correspond to its natural properties – mechanical, physical, chemical and others
2. the SM structure changes after the external influence energy W reaches the critical value exceeding the activation energy of the material restructuring process: $W > W_a$

3. the external influence energy is consumed to restructure the SM, but not the energy generated by the engineering system in which the material operates
4. the SM is capable of comparing the external influence energy with the extent of primary material restructuring; the result of this comparison is implemented in the feedback signal targeting the energy of external influence at secondary adjustment of material structure
5. the feedback is the main distinctive feature of self-organizing SM which is atypical for any other category of materials

When characterizing the SM, the term 'capability' was commonly used for individual features of living creatures as the subjective criterion of their successful existence. The synonymous notions 'properties' or 'capacity' are used in the materials science. The use of combination 'SM capability' is justified by the fact that the SM as the cybernetic system displays features of artificial intelligence identical to natural intelligence – the capability of living creatures thinking.

The SM regenerates after the external influence discontinues, in other words, it restores its characteristics. However the SM never regenerates absolutely due to the energy loss for restructuring (Joule heat emission, hysteresis, friction, incomplete phase transitions, etc.). The efficiency of SM restructuring $\eta = W_a/W < 1$, where W and W_a are energies of external influence and restructuring activation. The SM is an open thermodynamic system, which exchanges matter and energy with environment. Therefore, according to the first principle of thermodynamics, the classic relationship is justified for the SM:

$$dU - TdS - \delta A \leq 0,$$

where dU is internal energy change in SM under external effect, δS – the change in its entropy, δA – the work of SM restructuring, T – thermodynamic temperature. The sign of equality corresponds to the absolute inverse process of restructuring taking place in perfect SM.

According to the second principle of thermodynamics and from the standpoint of statistical physics, the SM as the system consisting of a very large number n of chaotically moving particles tends to regeneration, in other words, to spontaneous transition between states less probable into the states more probable. The second principle of thermodynamics is fulfilled with the larger probability the larger the n, and it has practically the nature of validity for the SM. At the same time, fluctuations take place continuously in the components of SM structure containing a small number of particles, that is, deviations occur from the second principles of thermodynamics.

A natural *direct feed* exists between the intensity of external influence on the material and the extent of its structural transformation. *Feedback* is the effect of results of some process on its evolution. The SM feedback is the process of mutual influence of external energy effect and the internal energy of transformed material structure. It determines the kinetics and the result of physical and chemical restructuring, that is, its finite state. In other words, the feedback is the form of mutual influence of the SM true state and variable conditions of its operation. The feedback is termed positive if the SM structure transforms not enough under the direct feed effect and the feedback accelerates this process. It is the negative feedback when the SM restructuring process changes the direction stabilizing the material structure. Thermodynamically expedient response to the variable primary SM reaction presumes the feedback flexibility (Wang 1998). There are mechanical, optical, electrical and other feedback types differing in the nature of processes using which the feedback is implemented in the SM. The SM feedback circuit may contain one or several links (structural unities of material) transforming the output signal by the set algorithm. The availability of feedback is the criterion of material intelligence.

Smart engineering system like the SM contains the feedback. However, it is not incorporated into the structure of a single material; its functions are performed by one or several components of the engineering system separated in space from the component manifesting the attributes of artificial intelligence. Examples of smart engineering systems are: the amplifier with the loud speaker emitting the sound signal affecting the microphone connected to the amplifier input; the friction unit implementing the effect of wearlessness; the system of electrochemical pipeline protection from corrosion.

The model of smart engineering system developed in the late 1940s by the designers of radio electronics and later assumed as the main principle of constructing the SM structure (Wang 1998) assumes interaction between three components – sensor, actuator and processor (from *sense* – sensation, *actuate* – to put into motion, *process* – to work up). The dissertation (Motyl 2000) proposes the form of their interaction in the SM structure shown in Fig. 1.1. The sensor is the sensitive component evaluating the nature and intensity of external effects on the material. The actuator is the executive component transforming the external effect into the material structure modification. The processor is the component controlling the actuator with signals from the sensor.

We have interpreted the interactions between these components of the material structure in the following way (Goldade et al 2005). Material *1* containing the sensor and actuator is the usual engineering material. Material *2* has the structural components acting as actuator and processor, while material *3* has structural components acting as sensor and processor. Both these materials can be active, in other words, be programmed to changes in effects of environment, or adaptive – compliant to the variable operating conditions. Smart material *4* contains the components

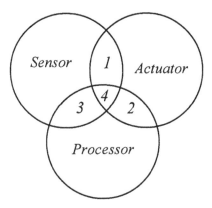

Fig. 1.1. SM structure model

acting as sensor, actuator and processor. Its structure changes expediently according to operating conditions; it affects the environment physically, chemically or biologically and, in addition, adjusts its own activity. In other words, it changes the level of responses depending on the operating conditions.

It is an abstract model and should be supplemented with characteristics of materials and physical and chemical processes induced by external effects.

Algorithm (from *algorithmi* – Latin transliteration of al-Chorizema – the name of an Arab mathematician of the 9th century) means the exact description of the method of the solution of the problem yielding the result determined unambiguously by the initial data. The SM mathematical modeling implies the derivation of algorithm controlling the material structure using the feedback system. It optimizes the value of external influence of energy expended on secondary restructuring following the criterion of compliance between SM operating properties and external effect intensity. In the theory of algorithms as a part of mathematics, there is a tendency to reduce the number of elementary operations and make them seem small. The algorithmic languages resulting from this approach are practically inapplicable to SM description. That is why the languages are developed in the theory of algorithms particularly oriented at these or other SM classes (Brebbia and Samartin 2000).

System approach is the trend of scientific cognition methodology implying consideration of the objects of study as systems. SM is a material cybernetic system and its model is a corresponding abstract cybernetic system. Any material is a system of interrelated structural components which conditions change under the effect of operating factors. The feedback signals affect the natural changes of the SM structural components additionally. The regularities of SM functioning are described by three families of functions (Glushkov 1973): i) governing the changes of structural components, ii) describing the material properties in response to the structural state, iii) setting the external effects on material. The SM is called deterministic if these functions are unequivocal; if the part of them is random functions, the SM is the probabilistic or stochastic system. The full SM description comprises additional a family of functions of material original state.

The mathematical apparatus best suitable for SM analysis is selected depending on discreteness or continuity of functions describing the SM as a system. It is the theory of differential equations for continuous functions, for discrete functions it is the theory of algorithms and the theory of automatons. The basic mathematical theory which is used in both cases, is the theory of information.

Types of structure controlling of materials can be the following.

The simplest is the programmable controlling: the operating conditions are assigned to material as the sequence of controlling effects (temperature, pressure, field, and environment), the feedback is absent. The material structure changes naturally under the effect of direct feedback. All materials operate under this program, only the smart ones are added with feedback action.

The simple controlling type is the classic self-control targeted at maintaining constant one or several parameters of the material structure using the feedback. An example is the system of automatic adjustment of corrosion inhibitor concentration in the volume of film package for metallic tools. To prevent corrosion, the wrapper should contain inhibitor vapors of some constant partial pressure. The inhibitor evaporates the wrapper volume inwards from the active polymeric film containing corrosion inhibitor. The evaporation discontinues when the evaporation rate becomes equal to the rate of condensation of inhibitor vapors on the tool surface. The evaporation resumes when the pressure declines to a critical value even when the vapors leak from loose wrapper (Pinchuk and Neverov 1993). The partial pressure of inhibitor vapors within the wrapper performs the function of feedback signal.

The aim of optimal controlling is to maintain the set value of some function from two groups of parameters. The parameters of the first group (external conditions) change spontaneously, while the parameters of the second group are adjusted by the signals of the controlling system. An example of the system of optimal controlling is protection of metallic articles from corrosion when they are not packaged tightly and then the pills of composite material used are based on polymer and volatile corrosion inhibitor. The polymeric matrix limits the rate of inhibitor evaporation from the pill. The partial pressure of inhibitor vapors delaying the metal corrosion depends on the level of humidity in the packing. The humidity varies in response to atmospheric conditions. It is necessary that the inhibitor evaporation rate changes in response to humidity variations. The task of optimal controlling of polymeric matrix structure restricting the inhibitor evaporation in response to humidity should be solved by the SM. In the simplest case, the task of optimal controlling comes to the maintenance of largest inhibitor vapor pressure within the packaging at the threshold value of humidity.

The problem of extremal controlling is solved, for example, by the system of notification of bridge overloading. The number of disturbing effects on its design (people, transport) is very great and their distribution along the bridge span is unpredictable. The smart component of this system can be a magnetostrictive material. The bearing components and bridge bearing structures fabricated from magnetostrictive alloys serve at the same time as pressure sensors. Under mechanical loading magnetic fields are generated in the structural elements, which inform about the excess of the permissible stress level. Strictly speaking, this system is not any adjustment system; it is rather an information system. In addition, the magnetostrictive sensors are made of material that can be referred rather to multifunctional than to smart materials. Nevertheless, this system reveals the signs of artificial intelligence. The feedback function missing in the magnetostrictive alloy is performed by other components of the tracking system of the overloading bridge.

If the uncontrollable SM operating conditions change substantially, the controlling system should maintain such material states which assure the controlling stability. When the effect of disturbing factors makes it impossible, SM self-tuning and self-organization are necessary. In case the engineering system is disturbed from equilibrium, the controlling system must change the parameters of links between structural components of SM to achieve controlling stability. Such controlling systems are called superstable. It can be assumed that most of SM will be superstable in the 21st century.

Some SM are capable of accumulating information about its structure state and in response to this, initiate processes of its restructuring. This SM property is called memory by analogy with the human brain. 'Memorizing' of information by SM takes place by changing states of structural components. It is a classic case of SM self-tuning. It may be illustrated by 'shape-memory-effect' or restoration of the original article shape by heating after plastic deformation.

The above examples prove that the engineering system may manifest attributes of artificial intelligence both being the SM and when it incorporates the multifunctional, adaptive or active materials. The lacking feedback function is performed by the controlling components of the engineering system.

An essential factor of SM efficiency (like all engineering materials) is reliability – the property of material to fulfill the assigned functions in the engineering system assuring the self-tuning of working parameters in response to operation conditions (at the assigned conditions of use and maintenance). The reliability assures the feasibility of SM application and the article it incorporates at the necessary time with required efficiency. The theory of reliability of cybernetic systems permits one to estimate the SM operating properties if their structural components damage, the feedback links break and so on (Mezzane and Luk'yanchuk 2008).

Methodology of SM study as cybernetic systems comprises three principally unlike methods. Two of them (experimental and theoretical) are broadly used in material science. The experimental method implies testing of the SM or its physical model. The essence of theoretical method is to describe the SM using mathematical apparatus followed by extraction of findings from this description, for instance, by solving accompanying system of equations. The third method termed as mathematical modeling, machine or virtual experiment, is one of the most essential results of cybernetic approach to study materials. The essence is that instead of testing real material, the experiment is carrying out with its mathematical description. The latter, together with the relevant software implementing changes in material characteristics, is entered into computer memory. It enables one to predict changes in properties under certain conditions, to change computer algorithms, of implementing properties, to display the three-dimensional structure, to calculate the parameters of physical properties and other factors (Sacellarion et al 2007). The processing speed of modern computers enables one to study SM structural transformation much faster than it occurs in reality. The method of virtual experiment has turned out most fruitful when SM variation is chosen among others which meet mostly to operating conditions.

Sometimes it is more appealing to learn the response to external effects than to explore the SM structure. In such a case, the modeling method implies application of the 'black box' principle mentioned in 1.1. The SM is likened to a 'black box' with unknown or too complicated structure to draw conclusions about the properties of composite material from the properties of its components.

1.3   Energy aspects of external influence

The materials are subjected to external actions of the interfaced system components and environment when incorporated into the engineering system. All external effects perform work on the material, alter its structure and internal energy. The internal energy is the sum of kinetic energy of molecules and atomic particles, which make up the material, plus the energy of interaction between all structural components. The internal energy is the function of material thermodynamic parameters (volume, temperature and others) the change of which is determined by the work of external effects. These effects can be physical, chemical or biological. The summarized characteristics of operating actions are listed below, which determine the reaction of materials, serve as energy source and determine feedback mechanisms.

The effect of physical nature is exerted on materials in the form of mechanical stresses, temperature and physical fields (that is fields of nuclear forces, gravitation and electromagnetic fields).

The mechanical stresses initiate collisions and responses of interaction between gas particles, homogenization or continiuty disruption of liquids, elastic and plastic deformation of solids. The deformation causes the restructuring of solid material, generation of isotropy or anisotropy, the extreme case being the disruption of continuity of the solid body. It begins with cracking and evolution of cracks and ends with separation of the solid specimen into parts. The mechanical stresses reinforced by friction cause fatigue of materials, which appears as local evolution of density of dislocations and vacancies. Tribochemical reactions are initiated on the surface and subsurface layers of the contacting bodies are destroyed. Thus, the energy of mechanical effect initiates the primary restructuring of the materials, which signals in the SM the feedback actuation.

The feedback mechanism is determined to a considerable extent by changes in internal energy resulting from structural transformation. According to the first principle of thermodynamics, the transition of adiabatically isolated material (receiving no outside heat) from state 1 to state 2 changes the internal energy U which is equal to the work PdV affected by the external mechanical force on the material during infinitely slow quasistatic process:

$$U_2 - U_1 = \int_1^2 PdV,$$

where P is pressure, V – volume of material. In the general case, according to the second principle of thermodynamics

$$dU = TdS - PdV,$$

where T is temperature, S – material entropy. The minimal U determines materials stable equilibrium at its constant entropy, volume and mass. Any deviations from the minimum are potential source of feedback energy in the SM.

The temperature is one of the main thermodynamic parameters of state characterizing the material thermal equilibrium. The temperature of any part of material is the same when it is in the state of thermodynamic equilibrium. The equilibrium temperature characterizes the intensity of thermal motion of atoms, molecules and other structural components.

The motion intensifies and diminishes when the temperature grows or lowers correspondingly. Variations of the intensity of thermal motion are due to phase transitions in materials – evaporation, distillation, crystallization, melting, condensation, and deviations from the diffusion equilibrium. At the cryogenic temperature (below 120 K), the plasticity (the ability of plastic deformation) and viscosity (the ability to absorb energy of external forces) of solid materials diminish. The processes evolving at non-equilibrium temperature (phase transitions, diffusion, viscosity variations, thermoelectric and other phenomena) induce changes in the internal energy of the material. They are studied by the thermodynamics of non-equilibrium processes. To determine their rate, the equations of balance of mass, impulse, energy and entropy in the material elementary volume are deduced and analyzed togther with the equation of the process in question. The phenomenological description of non-equilibrium process in the material consisting of n components is the following (Zubarev 1998):

$$J_i = \sum_{k=1}^{n} L_{ik} X_k ,$$

where J_i is the energy flux; index i characterizes energy type: heat, diffusive or impulse fluxes (the latest determines viscosity), electric current; L_{ik} – kinetic coefficient, X_k – thermodynamic force. The energy flux determines the feedback signal intensity in the SM.

Physical fields dictate interaction between structural components of the material. Nuclear forces acting between nucleons are attributed to the category of strong interactions and govern the structure and properties of nuclei of chemical elements. They are essential for carrying out smart nuclear reactions. The gravitation field is the universal interaction between all types of matter. The effect of this compulsory satellite of all processes evolving in materials should be taken into account particularly when assessing the state of colloid systems in liquid and gaseous phases. The electromagnetic field dictates interactions between electrically charged particles or particles possessing the magnetic momentum. Particular electromagnetic field cases are the electric field created by immobile charge carriers, the magnetic field appearing around immobile conductors of direct current as well as around constant magnets. The external electromagnetic field interacts with charged particles generating induction current. The conductivity currents dictate energy losses and attenuation of electromagnetic oscillations in the material. The currents can induce the electric polarization and magnetize the material, altering its dielectric and magnetic permeability. The electromagnetic oscillations (waves) with the spectrum of radio waves, light, X-ray and γ-emission propagate in materials with limiting velocity depending on the wavelength (in range from 10^{-15} to 10^3 m). Polarized and magnetized materials are sources of electric and magnetic fields. If the material is homogenous and isotropic, then according to the Maxwell theory, under the effect of planar monochromatic electromagnetic waves, the intensity of electric E and magnetic H fields corresponds to the following equations (Miguilin 1998):

$$E = E_0 \cos (kr - \omega t + \varphi);$$

$$H = H_0 \cos (kr - \omega t + \varphi),$$

where E_0 and H_0 are amplitudes of oscillations of electric and magnetic fields, $\omega = 2\pi\nu$ – circular frequency of these oscillations, φ – random phase shift, k – wave vector, r – radius-vector of point, $\mathbf{E} \perp \mathbf{H} \perp \mathbf{k}$, $H_0 = \sqrt{\dfrac{\varepsilon}{\mu}} E_0$, ε and μ – dielectric and magnetic permeability. The energy of generated fields is used for feedback functioning.

The chemical effect on materials exerts the media in which materials operate. Chemical resistant materials do not react chemically to the media, but absorb active molecules from the medium, which concentrate in the surface layer of condensed (solid and liquid) materials. This phenomenon of adsorption alters the composition of surface layer and the surface energy of the material. There are physical adsorption and chemical adsorption depending on the nature of links at the interface material/adsorbate. The physical adsorption is due to the Van der Waals forces, which do not induce any significant changes in the electronic structure pf adsorbed molecules. The chemical adsorption (chemisorption) is accompanied by formation of chemical bonds; it is a chemical reaction evolving in the area limited by the surface layer (Dubinin and Serpinskii 1998). The fundamental thermodynamic equation of adsorption of Gibbs is the following:

$$d\sigma + \sum_i \frac{n_i^\omega}{s} \cdot d\mu_i = 0,$$

where σ is interphase surface tension, S – interface surface area, μ_i and n_i – chemical potential and number of moles of i-th component of adsorption system (Zubarev 1998). The adsorption directly affects the surface energy of materials and consequently – the feedback formation mechanisms in the SM.

The chemical reactions transform the components into other products, without altering the nuclei of atoms. These transformations require definite conditions – temperature, pressure, irradiation and others. Heat generation, illumination, alterations of the aggregate state and other energy transformations, which are expedient for actuating feedback links, may accompany the chemical reactions. The energy barrier value between the initial and final states of reacting material is called the activation energy.

It determines the change ΔG in the Gibbs energy. The required condition of spontaneous (without energy input from outside) reaction is to reduce the Gibbs energy: $\Delta G < 0$. The overwhelming majority of chemical reactions is irreversible, in other words, alongside with direct transformation of reagents into products, an inverse reaction takes place. If the direct and inverse reactions evolve with the same rate, the chemical equilibrium is achieved. For the reaction

$$\sum_i v_i A_i \rightleftharpoons \sum_j v_j B_j,$$

where A_i are original reagents, B_j– products, v_i and v_j – their stechiometric coefficients, the chemical equilibrium is achieved at the condition:

$$\sum_j v_j \mu_j - \sum_i v_i \mu_i = 0,$$

where μ_i and μ_j are chemical potential of original reagents and reaction products. The local chemical equilibrium is achieved in smaller structural component if the material has a non-equilibrium structure. It corresponds to local values of temperature, pressure, chemical potentials of structural components (Zubarev 1998).

The probability of material involvement in the chemical reaction and its direction depend on the thermodynamic (entropy, ΔG) and kinetic factors (activation energy, the value of the pre-exponential multiplier in the Arrenius equation). Depending on the method of initiation of the active state of material and environment, there are plasma-chemical, radiation-chemical, thermal, photo-thermal, and electrochemical reactions. The variety of energy sources of reactions activating dictates the great number of energy techniques of feedback stimulation.

The biological effect on materials is rendered primarily by microorganisms (bacteria, microscopic fungi, sometimes – the animalcular and viruses) organized into biologic systems (biosystems). The biosystem is the combination of interrelated and interacting microorganisms, the properties of which are the sum of properties of the organisms in it. The biosystem is a smart system capable of adapting itself to the environment, evolving, reproducing and developing. The biosystem is an open system exchanging energy, matter and information with the environment. This exchange is based on the genetic information under control of specific controlling mechanisms adjusting the use of energy from environment by the biosystem. The stability of stationary states of the microorganisms of the biosystem (constant internal characteristics under environment changes), as well as the instability of stationary states (the ability to transit from one to another state) are assured by the variety of self-regulation mechanisms of the biosystem. The self-regulation is based on the feedback principle, positive or negative. The biosystem transition from one state to another is the method of its adaptation to changing external conditions (Giliarov 1998).

The cause of the influence of the biosystem on the engineering materials is the need for energy resources to assure vital functions of microorganisms. Materials play the role of energy sources. The enzymes, or the biological catalysts, which are proteins by chemical nature, play the main role at the molecular level when the biosystem and material interact. The enzymes catalize the conversion of internal energy by microorganisms, first, into the energy accumulated by the ATP (adenosine triphosphate, presents in all living cells), and second, into the energy of transmembrane electrochemical potentials. The latter appear due to the gradients of ions H^+ or Na^+ concentration on different sides of biological membrane. The accumulated energy is used to synthesize the ATP, to move microorganisms, for active transport in the cells of ions, carbohydrates and amino acids (Skulachev 1998).

The biological damage of materials occurs due to the biochemical reactions initiated by the mold fungi and oxidizing bacteria (able to live in anaerobic conditions, without oxygen). It intensified disastrously during the last half-century owing to the industrial pollution of the environment (Bott 2011). The microbiological corrosion results from oxidizing fermentation in the presence of oxygen and the reductaze accelerating the disintegration of hydrocarbons. The aggressive products of this catalytic process are organic acids corroding metals and organic materials (Blagnik and Zanova 1965).

The anabolic reactions (aimed at renewing cells) are the basis of biosynthesis or formation of organic compounds from simple substances. The biosynthesis is an essential component of metabolism among all living creatures. The biosynthesis mechanism is determined by the feedback, or hereditary information encoded in the cell genetic apparatus. The biosynthesis serves as the technology of industrial production of vitamins, some hormones, antibiotics, amino acids, and fodder proteins.

To conclude the present paragraph, it is worthwhile to note that the energy influence of operating factors on the material dictates the natural structural transformation and alteration of the internal energy. This material 'response' to external effects generates the prerequisites for correction its impaired state using the feedback. It is generated from the energy of external influence absorbed by the material targeting a portion of it for activation physical and chemical processes of secondary restructuring. The next paragraph deals with the main regularities of these processes mailing the material filling of the abstract model 'sensor-actuator-processor'.

1.4   Systematization and methods of smart materials description

To represent any group of materials, the obligatory information should comprise a summarized characteristic and location of this group in the system of subordinated groups in the whole totality of materials. The preceding paragraphs list sufficient attributes classifying the SM, but none of them is exhaustive. The SM system is better represented using the ideas of typology as the method is based on the division of the combination of objects and then regrouping them using the generalized model or type.

The SM model is represented below incorporating the main components: physical, mechanical, chemical and biological processes evolving in the material in operation. They determine the mechanism of secondary restructuring which transform the material under the feedback control. The formalization of these components with the methods of mechanics permits to obtain the summarized description of interactions in the structure of smart materials and engineering systems.

The mechanism of physicochemical action of the feedback was chosen as the attribute of SM taxonomic table construction. Representation of the SM combination with the taxonomic methods attracts by the possibility to determine promising (at the present level of natural and engineering sciences) means of the structure and properties of materials controlling, to discover the effective and expedient ways of novel SM development.

1.4.1   Methods of materials taxonomy

In the second half of the 20th century, the general tendency of scientific evolution, including materials science, has enhanced of the role of typology in scientific thinking. The advent of computers has accelerated the development of materials with specific properties using the typology to choose promising structural modification of materials, optimal technologies of their fabrication and processing.

The term 'taxonomy' proposed in 1913 by Swiss botanist A. Decandolle was used for a long time in biology as the synonym of systematics. The notion taxonomy in 1960–70s narrowed to one of the parts of systematics as the method of categorizing subordinated groups of objects or *taxons*. That made it easier to classify all objects without losing the information content, avoid contradictions and reflect the reality objectively. The first endeavor to systematize materials with the methods of taxonomy was undertaken by Russian materials scientist S.M. Brekhovskikh in the 1970s (Breshkovskikh 1981).

The analysis of engineering materials with the categories of taxonomy implies the following. The criterion of systematization is the combination of three attributes: the conditions of application, determinative properties, and structure of materials.

The conditions of application of materials (A) are subdivided into the following hierarchic levels: types (A_b), classes $(A_{b,c})$ and kinds $(A_{b,c,d})$. The division is based on subsequently specified characteristics of attribute A in the main condition of application (a, b, c) or in the additional condition (i, g, h). The types of conditions of application of material cover time, temperature, environment, electric and magnetic fields, electromagnetic and corpuscular emissions, pressure, acceleration and other factors.

The determinative properties of materials (B), by subsequent specification of the characteristics of properties f, j, k, are also subdivided into types (B_f), classes $(B_{f,j})$ and kinds $(B_{f,j,k})$. The types of determinative properties cover mechanical, electrical, magnetic, optical and other properties of materials.

The material structure (C) is analyzed at three levels: macrostructure (P_n^r), microstructure (P_m^e) and atomic-molecular structure (P_j^x). Within each level the structure is characterized by its components, their mutual arrangement and nature of links between components. The general structure of each material (C^M) can be described by the function depending on 9 parameters:

$$C^M = C^M [C^r(P_n^r, L_e^r, M_d^r), C^e(P_m^e, L_k^e, M_s^e), C^x(P_l^x, L_f^x, M_j^x)],$$

where P_n^r, P_m^e, P_l^x are the components of macro-, micro- and atomic-molecular structures; L_e^r, L_k^e, L_f^x – arrangement of components in macro-, micro- and atomic-molecular structures; M_d^r, M_s^e, M_j^x – types of links between components in macro-, micro- and atomic-molecular structures.

The group of materials combined by common attributes A, B and C, form the taxon. Based on the above principles and using the determinant attributes of the taxon, one can classify materials, to estimate the potential of materials science and set up a computerized information system for materials search (Breshkovskikh 1981).

The estimation of the potential of materials science is first of all to perform structural and functional analysis targeted at the search of non-existent but possible engineering structures of materials. By analyzing the system of subordinated taxons combined by common structural parameters, S.M. Brekhovskikh forecast the properties of some materials now unavailable. The diagram of three-dimensional computerized data system of materials in which the elementary data bits are taxonomic categories A, B and C of materials, was given in textbook (Pinchuk et al 1989). The system can accumulate information, analyze it with software methods and compare the data requested by consumers and designers. New possibilities of handling these databases belonging to different researchers, centers and states open up due to information globalization. Thus, the Internet technologies stimulate the unification and standardization of materials according to taxonomic principles.

1.4.2   Smart material model

An unavoidable sequence of interaction between material and environment is the restructuring, which we term primary. The primary restructuring is the response of material to external effects resulting from physical,
chemical or biological processes of degradation. The restructuring mechanisms and its kinetic regularities are individual for each material; they
depend on the nature and intensity of the external effect. For instance,
the amplitude of thermal oscillations of atoms is approximately inversely
proportional to the mass of atoms and to the forces of chemical bonding
between atoms in the crystalline structure, and directly proportional to
the material temperature. In the first approximation of spherically symmetric (isotropic) oscillations, the probability $\omega(r)$ that the atomic centers
are located at the distance r from the perfect position is described with the
Gaussian distribution (Vanshtein 1990):

$$\omega(r) = (2\pi \bar{u}^2)^{-3/2}\exp(-r^2/2\bar{u}^2), \qquad (1.1)$$

where $\bar{u}^2$ is the atom mean quadratic displacement as the function of its
mass and external energy effect, links atoms, and material temperature.

It is understandable that natural laws are impossible to change and
compel atoms to perform thermal oscillations otherwise than equation
(1.1) determines. However, it is possible to redistribute the external energy
effect in the material so that the value $\bar{u}^2$ changes locally in different
material portions. The smart materials differ from simple ones by the
availibility of feedback system, which corrects the primary response of
the SM to the influence of the environment.

The final SM response to the external effect results in the smart correction of susceptability to the operating conditions. The physical and
chemical mechanisms perform the correction mainly like the primary
restructuring. The SM structural transformation, which would take place
due to external effect of energy controlled by feedback signals, is called
the *secondary restructuring* (Fig. 1.2). It first, stipulates 'saving' the internal
energy of the material extending its service life and, secondly, optimizes
the performance parameters on the criterion of minimal energy consumption. Due to this, the SM structure changes with time fooling the principle
'necessary and enough' under external effects of various nature. It com-
plies with the laws of energy and mass conservation; the latest is interpreted in modern formulation as the permanency of the sum of material
mass and the mass equivalent to the obtained or expended energy.

It is remarked in 1.2 that the feedback reinforcing the primary process is positive, that weakening them is negative. The positive feedback
usually renders the cybernetic system unstable; the negative feedback
moderates its functioning making the system more stable. Therefore, the
negative feedback in SM is preferable in most cases.

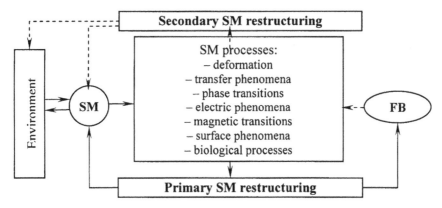

Fig. 1.2. Diagram of links of interactions between SM and environment implemented using feedback (FB) system. Full lines correspond to primary restructuring; dashed lines correspond to secondary restructuring

Let us exemplify it in the following way. The systems with a feedback are often represented (Krylov and Lukjanchuk 1992) in a diagram with the signal from the output of an amplifier coming in its input (Figure 1.3). The unit 'transducer' is a device implementing the conversion of the input signal z into the output signal Z according to the specified law which the next unit branches into its part – signal x.

The conversion $x \to X$ is performed in the feedback circuit following the typical algorithm. The complete description of the system in Figure 1.3 assumes the awareness of at least two rules: signal x in the feedback circuit is deducted from signal Z at the amplifier output, and signal X is added to signal U_{in} at the system input. The feedback characteristic is the coefficient of signal transmission β along the channel $X = \beta Z$. The systems with negative feedback have $U_{out} < U_{in}$, therefore $x < X < 0$ и $\beta = X/Z < 0$.

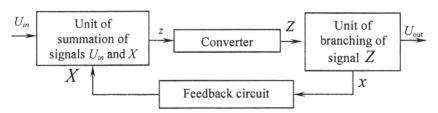

Fig. 1.3. Block-diagram of feedback system (Krylov and Lukjanchuk 1992)

The work performed by the SM during secondary restructuring under the feedback effect can be rated with the following considerations. Let us assume the SM structure as combination of material points with masses $m_1, m_2, ..., m_n$ oscillating with velocities $v_1, v_2, ..., v_n$. Assume $F'_1, F'_2, ..., F'_n$

– resultant interacting internal forces, $\mathbf{F}_1, \mathbf{F}_2,...,\mathbf{F}_n$ – resultant external conservative forces (their work is independent of the particle motion path, it depends on just the particle initial and finite positions) acting in the SM. Let us assume that each material point is affected also by external conservative forces with the resultant $\mathbf{f}_1, \mathbf{f}_2,...,\mathbf{f}_n$. According to the law of energy conservation, there is the following ratio between them (Trofimova 1990),

$$\sum_{i=1}^{n} m_i(\mathbf{v}_i d\mathbf{v}_i) - \sum_{i=1}^{n}\left(\mathbf{F}_i{'}+\mathbf{F}_i\right)d\mathbf{r}_i = \sum_{i=1}^{n} \mathbf{f}_i d\mathbf{r}_i, \qquad (1.2)$$

where $d\mathbf{r}_i$ is displacement of point i during time interval dt. The first term in the left equation part is

$$\sum_{i=1}^{n} m_i(\mathbf{v}_i d\mathbf{v}_i) = \sum_{i=1}^{n} d(m_i \mathbf{v}_i^2 /2) = d\mathbf{T},$$

representing the kinetic energy increment $d\mathbf{T}$ of SM particles. The second term $\sum_{i=1}^{n}(\mathbf{F}_i{'}+\mathbf{F}_i)d\mathbf{r}_i$ is equal to the work of internal and external conservative forces acting on SM particles. This work in the equation is designated with negative sign; hence, it is equal to the SM potential energy increment dP. The right term $\sum_{i=1}^{n} \mathbf{f}_i d\mathbf{r}_i$ of equation (1.2) characterizes the work A of external conservative forces acting on the SM. Thus, $d(T + P) = dA$.

The work is accomplished when the SM after the primary restructuring (in state 1) passes to state 2 which occurs after the secondary restructuring:

$$\int_{1}^{2} d(T + P) = A_{12}.$$

It means that changes in kinetic and potential energy during secondary SM restructuring are equal to the work expended by nonconservative external forces controlled by the feedback system. All materials (all systems in general in nature) are, strictly speaking, dissipative, in other words, their internal energy $(T + P)$ diminishes with time and converts into other energy forms (the aging of materials).

The diagram in Fig. 1.2 shows that the secondary cycle is virtually locked in the 'black box' where the SM undergoes the structural modification processes. The external effects are transmitted to the material indirectly and through the feedback system. The SM properties outputted by the 'black box' are optimized according the energy conservation criterion.

Hence, the SM model can be represented simplified as shown in Fig. 1.4. The comment is the following.

The environmental influence, in other words, the operating conditions, causes primary SM restructuring. The feedback system sums up the data about the degree of restructuring and the intensity of external effects and then initiates (by external energy) the SM secondary restructuring, which determines the SM operating properties. The transmission mechanism from the environment to the feedback system and also the energy, physical, chemical and biological mechanisms of secondary restructuring are, principally, understandable, though they remain unclear. That is why the SM structure represented in Fig. 1.4 as dashed rectangle can be likened to the 'black box' in which the evolving processes need further exploration.

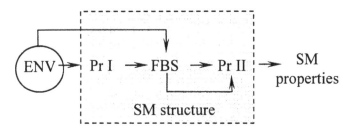

Fig. 1.4. Smart material simplified model: ENV – environment, Pr I and Pr II – processes of primary and secondary restructuring, FBS – feedback system

1.4.3   Classification of smart materials and engineering systems

The systems of SM classification in the first publications dealing with this class of materials were rather common. The SM were divided into groups by attributes traditionally used in the science of materials: by purpose (electrorheological and magnetic fluids, alloys with shape memory, fiber optic and others) (Wang 1998), by structure (crystalline, noncrystalline, atomic clusters, ligands and others) (Gandhi and Thompson 1992), by type of energy effects (electronic, magnetic, thermal and others) (Singh 2005). The ideas developed in the 1950s about smart radio electronic circuit (sensor – actuator – processor) were expanded a lot in the last decade: the mathematical models and methods of description of functions of triad components in the SM were proposed (Lazzari et al 2002), the ideas about interaction between components in mechanical systems were developed (Varadan et al 2006), methods of SM modeling were advanced with the account of physical and chemical nature of the feedback system and mobility of interphase boundaries (Laso 2009).

The feedback component in the diagram (Fig. 1.2) is divided in space from other circuit components. It presumes that the feedback can

Table 1.1 Taxonomic presentation of smart materials (SM) combination

Feedback nature	Mechanical					Tribological			Thermal					Emissive			Electric			Magnetic			Diffusive			Chemical					Biological	
Mechanisms of feedback effect on SM properties	Stress appearance	Restructuring	Specimen deformation	Rheological	Tribological	Lubrication	Surface layer restructuring	Internal friction modification	Thermal expansion	Restructuring	Melting	Solidification	Structural alteration	Melting	Chemical reactions	Emission of portion of radiation	Current appearance	Electric polarization	Alteration of conductivity type	Alteration of magnetic texture	Alteration of domain structure	Electric polarization	Concentrative diffusion	Thermal diffusion	Electric diffusion	Photochemical reactions	Radiation chemical reactions	Mechanochemical reactions	Electrode reactions	Plasmachemical reactions	Biocorrosion	Biosynthesis
Designations	B_1	B_2	B_3	B_4	B_5	B_6	B_7	B_8	B_9	B_{10}	B_{11}	B_{12}	B_{13}	B_{14}	B_{15}	B_{16}	B_{17}	B_{18}	B_{19}	B_{20}	B_{21}	B_{22}	B_{23}	B_{24}	B_{25}	B_{26}	B_{27}	B_{28}	B_{29}	B_{30}	B_{31}	B_{32}
Deformation A_1	A_1B_1	A_1B_2	A_1B_3	A_1B_4	A_1B_5	A_1B_6	A_1B_7	A_1B_8	A_1B_9	A_1B_{10}	A_1B_{11}	A_1B_{12}	A_1B_{13}	A_1B_{14}	A_1B_{15}	A_1B_{16}	A_1B_{17}	A_1B_{18}	A_1B_{19}	A_1B_{20}	A_1B_{21}	A_1B_{22}	A_1B_{23}	A_1B_{24}	A_1B_{25}	A_1B_{26}	A_1B_{27}	A_1B_{28}	A_1B_{29}	A_1B_{30}	A_1B_{31}	A_1B_{32}
Friction A_2	A_2B_1	A_2B_2	A_2B_3	A_2B_4	A_2B_5	A_2B_6	A_2B_7	A_2B_8	A_2B_9	A_2B_{10}	A_2B_{11}	A_2B_{12}	A_2B_{13}	A_2B_{14}	A_2B_{15}	A_2B_{16}	A_2B_{17}	A_2B_{18}	A_2B_{19}	A_2B_{20}	A_2B_{21}	A_2B_{22}	A_2B_{23}	A_2B_{24}	A_2B_{25}	A_2B_{26}	A_2B_{27}	A_2B_{28}	A_2B_{29}	A_2B_{30}	A_2B_{31}	A_2B_{32}
Temperature A_3	A_3B_1	A_3B_2	A_3B_3	A_3B_4	A_3B_5	A_3B_6	A_3B_7	A_3B_8	A_3B_9	A_3B_{10}	A_3B_{11}	A_3B_{12}	A_3B_{13}	A_3B_{14}	A_3B_{15}	A_3B_{16}	A_3B_{17}	A_3B_{18}	A_3B_{19}	A_3B_{20}	A_3B_{21}	A_3B_{22}	A_3B_{23}	A_3B_{24}	A_3B_{25}	A_3B_{26}	A_3B_{27}	A_3B_{28}	A_3B_{29}	A_3B_{30}	A_3B_{31}	A_3B_{32}
Emission A_4	A_4B_1	A_4B_2	A_4B_3	A_4B_4	A_4B_5	A_4B_6	A_4B_7	A_4B_8	A_4B_9	A_4B_{10}	A_4B_{11}	A_4B_{12}	A_4B_{13}	A_4B_{14}	A_4B_{15}	A_4B_{16}	A_4B_{17}	A_4B_{18}	A_4B_{19}	A_4B_{20}	A_4B_{21}	A_4B_{22}	A_4B_{23}	A_4B_{24}	A_4B_{25}	A_4B_{26}	A_4B_{27}	A_4B_{28}	A_4B_{29}	A_4B_{30}	A_4B_{31}	A_4B_{32}
Electric field A_5	A_5B_1	A_5B_2	A_5B_3	A_5B_4	A_5B_5	A_5B_6	A_5B_7	A_5B_8	A_5B_9	A_5B_{10}	A_5B_{11}	A_5B_{12}	A_5B_{13}	A_5B_{14}	A_5B_{15}	A_5B_{16}	A_5B_{17}	A_5B_{18}	A_5B_{19}	A_5B_{20}	A_5B_{21}	A_5B_{22}	A_5B_{23}	A_5B_{24}	A_5B_{25}	A_5B_{26}	A_5B_{27}	A_5B_{28}	A_5B_{29}	A_5B_{30}	A_5B_{31}	A_5B_{32}
Magnetic field A_6	A_6B_1	A_6B_2	A_6B_3	A_6B_4	A_6B_5	A_6B_6	A_6B_7	A_6B_8	A_6B_9	A_6B_{10}	A_6B_{11}	A_6B_{12}	A_6B_{13}	A_6B_{14}	A_6B_{15}	A_6B_{16}	A_6B_{17}	A_6B_{18}	A_6B_{19}	A_6B_{20}	A_6B_{21}	A_6B_{22}	A_6B_{23}	A_6B_{24}	A_6B_{25}	A_6B_{26}	A_6B_{27}	A_6B_{28}	A_6B_{29}	A_6B_{30}	A_6B_{31}	A_6B_{32}
Diffusion A_7	A_7B_1	A_7B_2	A_7B_3	A_7B_4	A_7B_5	A_7B_6	A_7B_7	A_7B_8	A_7B_9	A_7B_{10}	A_7B_{11}	A_7B_{12}	A_7B_{13}	A_7B_{14}	A_7B_{15}	A_7B_{16}	A_7B_{17}	A_7B_{18}	A_7B_{19}	A_7B_{20}	A_7B_{21}	A_7B_{22}	A_7B_{23}	A_7B_{24}	A_7B_{25}	A_7B_{26}	A_7B_{27}	A_7B_{28}	A_7B_{29}	A_7B_{30}	A_7B_{31}	A_7B_{32}
Chemical reaction A_8	A_8B_1	A_8B_2	A_8B_3	A_8B_4	A_8B_5	A_8B_6	A_8B_7	A_8B_8	A_8B_9	A_8B_{10}	A_8B_{11}	A_8B_{12}	A_8B_{13}	A_8B_{14}	A_8B_{15}	A_8B_{16}	A_8B_{17}	A_8B_{18}	A_8B_{19}	A_8B_{20}	A_8B_{21}	A_8B_{22}	A_8B_{23}	A_8B_{24}	A_8B_{25}	A_8B_{26}	A_8B_{27}	A_8B_{28}	A_8B_{29}	A_8B_{30}	A_8B_{31}	A_8B_{32}
Biological A_9	A_9B_1	A_9B_2	A_9B_3	A_9B_4	A_9B_5	A_9B_6	A_9B_7	A_9B_8	A_9B_9	A_9B_{10}	A_9B_{11}	A_9B_{12}	A_9B_{13}	A_9B_{14}	A_9B_{15}	A_9B_{16}	A_9B_{17}	A_9B_{18}	A_9B_{19}	A_9B_{20}	A_9B_{21}	A_9B_{22}	A_9B_{23}	A_9B_{24}	A_9B_{25}	A_9B_{26}	A_9B_{27}	A_9B_{28}	A_9B_{29}	A_9B_{30}	A_9B_{31}	A_9B_{32}

Note: light squares correspond to available SM and smart engineering systems, dark squares correspond to be developed materials and systems be integrated into the SM structure or its functions can be executed by other components of the smart engineering system. It can be assumed the model in Fig. 1.2 characterizes both the SM and smart engineering systems. We divided the integrity of these objects into taxons 'feedback effect – feedback circuit mechanism'. The criteria of environment type selection are rather contradictory: on the one hand, it is desirable to include the effect of all fundamental influence in material systems, on the other hand, only those which are most significant in modern engineering should be selected. The same criteria were applied to the feedback systems. When detailing the feedback actuation mechanisms, state-of-the-art methods were preferred which are most typical for correcting primary responses to external effects. This question also remains debatable because the feedback detailing extent can be virtually endless.

From the considerable of keeping the book in reasonable scope, we have considered only 9 types of environmental effects of physical, chemical and biological nature and designated them with symbols $A_1 \div A_9$, and 32 main mechanisms of feedback circuit implementation designated $B_1 \div B_{32}$. It permits us to classify SM and smart engineering systems by the criterion of feedback, physical and chemical nature and for the first time to represent their totality as taxonomic Table 1.1. The light squares correspond to the available SM, the dark squares to those to be developed. It is provisionally enough due to a simple reason: nobody possesses all information about SM because each moment a novel SM may appear unknown heretofore, which will be reported in the press or the internet afterwards.

Let us briefly characterize the taxons in Table 1.1.

1. **Taxons with feedback system acting by deformation mechanisms.**
   The feedback system controls the distribution of mechanical stresses (tension-compression, bending, torsion and others). When they grow above the critical limit, the liquid components flow, solid components deform and change the structure. The restructuring occurs in the volume, on the phase interface, or in the surface layer of SM and may be accompanied by physical and chemical effects (mechanostriction, mechanoelectret appearance, mechanocaloric effect and others). The rheological mechanism of the SM state controlling initiates diffusion, creep, stress relaxation etc. The friction mechanism of feedback functioning induces squeezing out the lubricating fluid from the SM, generates internal friction, heat and other processes.

   In the 1990s, the need to control the stress state of materials throughout their life time (Paton and Nedoseka 1994) became evident. This challenge became global in the 21st century when the database and facility of nondestructive diagnostics have grown immeasurably (Prokhorenko 2003). The last achievements of the physics of nondestructive control have provided an inexhaustible base for feedback cicuits development, which optimize the responses of materials to operating conditions by deformation.

2. **Taxons with feedback system of tribological nature.** The feedback problem in this case is optimization in the first place: the friction unit lubricating conditions in which the material is applied; the surface layer structure of a SM sample movably contacting the counter-body; the internal friction parameters in the SM.

   During friction the sheer stresses localize in surface layers of the rubbing samples. It of makes the performances of the friction unit stronger than the volume properties of materials. Since the materials operate in tribological systems, the feedback system does not have to be incorporated into the SM. Its functions can be performed by other components. The significance of the global problem of friction should be emphasized here. The machinery friction consumes a huge amount of generated power and it makes quite a burden for mankind. Even a negligible decrease of power loss through friction by the use of SM would much improve the efficiency of engineering devices.

3. **Taxons with feedback system exerting thermal influence.** The temperature drops in materials are due to heat input-output, generation of heat by friction, restructuring of materials accompanied by exo- and endothermic effects. The feedback system optimizes the operating state of materials giving rise to:
   – thermal expansion;
   – restructuring of solid phase (transitions of order II);
   – melting of crystalline materials, transition of vitreous materials into highly elastic and fluid state, thermal disordering of liquid crystalline structure;
   – solidification or, that is to say, fluid transition into solid state, highly elastic and highly fluid into vitreous state and the homogeneous fluid into liquid crystalline state. The latter transition relates to the feedback group conditionally in order to reduce the number of taxons of feedback mechanisms.
   The operation of materials at temperature drops is a rule in modern engineering, and rare exceptions virtually do not influence the choice of materials. The thermal restructuring and the phase state create favorable prerequisites for SM developing.

4. **Taxons with feedback system acting by emission.** The physical effects of emission interaction with matter can be used in the feedback systems being the basis of numerous studies of the structure, physical and chemical modification of materials. The emission

energy of feedback can alter the structure or melt solid materials, initiate the physical and chemical interaction between components, generate emission of particles and radiation.

5. **Taxons with feedback system of electric nature.** The feedback system maintains performance of materials inducing electric current, electric polarization, alteration of conductivity type and other effects. The feedback of electric nature permits to concentrate a considerable quantity of energy in structural micro-components, easily distribute it through the specimen volume and convert electric energy into other types.

6. **Taxons with feedback system of magnetic nature.** The external effects generate magnetic field in some materials. The feedback system expends this energy to compensate structural damages occurring on materials operation. In this case, the feedback initiates the restructuring of magnetic texture or domain structure or converts them into the magnetoelectret state.

   The achievements in the study of magnetism permitted to create naturally absent materials – ferrites for super high frequency devices, highly coercive compounds like $SmCo_5$, amorphous magnetics, transparent ferromagnetics and others. Their application in the feedback structure had made for development of cybernetic systems maintaining the assigned level of operating properties throughout the whole life time of materials for radio-electronic equipment.

7. **Taxons with diffusive feedback system.** The material operation is accompanied by the diffusion of environment components or the diffusion of materials components. The feedback directs the energy of external effects to maintain the assigned material performance by initiating the concentration diffusion, thermo-, electro- and other types of diffusion at certain parts of material.

   The diffusion limits the chemical reactions setting the rate of evaporation, crystallization, condensation and dissolution of crystals. It plays an important role in the life of biological cells. It proves a reach potential of diffusive feedback in SM, in smart technological systems and medicinal articles.

8. **Taxons with feedback system controlling chemical reactions.** The feedback transforms the energy of external effect on the material into the activation energy of chemical reactions between the material and environment or between components of the material.

   The chemical interactions create broad opportunities for feedback system implementation. The reactions of decomposition, oxidation, polymerization and others flow in different ways depending on the stehiomectic ratio between components, the extent of their conversion, constants of velocity and equilibrium, activation energies, thermal effect. The feedback system heals the operating damages of material structure initiating photochemical, radiation-chemical, mechanochemical, plasma-chemical reactions or electrode processes.

9. **Taxons with feedback system of biochemical nature.** These SM operate in the biological environment which life activity products damage (cause degradation) of the structure and impair the material properties. The feedback system directs the energy of external biochemical effect at retarding the biological corrosion of materials and adjustment of their physical and chemical interaction with the components of biological environment involved in biosynthesis.

The biocorrosion results from chemical reactions of the material with intermediate or final products of metabolism – metabolites. The feedback system inhibits the physical and chemical activity of metabolites and localizes the reactions with them over certain volumes of the material. The biosynthesis products, or active substances produced by biotechnological processes, influence materials strongly. The feedback system purposefully alters the flow of reactions between biosynthesis products and components of material structure.

Thus, the model of SM and smart engineering system as cybernetic objects is proposed in which the physical and chemical processes running in the materials on operation are elements of this model. A part of the energy of these processes the feedback system and turns into the secondary restructuring of materials prolonging the useful operating life of the objects. The taxonomic table of SM and smart engineering system is developed based on the model and using the ideas of materials mechanics. The physical and chemical mechanisms of feedback action are selected as the distinctive attribute of taxons.

The SM 'filling' taxons in Table 1.1 belong to the different classes of materials: of electric engineering, electronic and medical devices, structural, triboengineering, protective, packaging and biologically active materials. The examples of SM and smart engineering systems of these classes corresponding to the white squares in the Table are listed in the next chapter.

References

Blagnik, R. and V. Zanova. 1965. *Microbiological Corrosion*. Chemistry, Leningrad.
Bott, T. Reg. 2011. *Industrial Biofouling*. Elsevier, London.

Brebbia, C.A. and A. Samartin. 2000. *Computational Methods for Smart Structures and Materials*. WIT Press, Southampton.

Breshkovskikh, S.M. 1981. Scientific principles of systematic of materials. pp. 63–75. *In*: Scientific Principles of Materials Science. Nauka, Moscow.

Dubinin, M.M. and V.V. Serpinskii. 1998. Adsorption. pp. 53–63. *In*: I.L. Knunianc (ed.). *Chemical Encyclopedia* in 5 volumes, Vol. 1. Soviet Encyclopedia, Moscow.

Gandhi, M.V. and B.S. Thompson. 1992. *Smart Materials and Structures*. Springer, Berlin.

Giliarov, M.S. (ed.). 1998. "Biological systems". p. 65. In: *Biology: Big Encyclopedic Dictionary*. Soviet Encyclopedia, Moscow.

Glushkov, V.M. 1973. "Cybernetics". pp. 211-225. In: A.M. Prokhorov (ed.). *Big Soviet Enciclopedia in 30 volumes*, Vol. 12. Soviet Encyclopedia, Moscow.

Goldade, V.A., L.S. Pinchuk, A.V. Makarevich and V.N. Kestelman. 2005. *Plastics for Corrosion Inhibition*. Springer-Verlag, Berlin-Heidelberg.

Krylov, N.V. and B.S. Lukjanchuk. 1992. "Feed-back". pp. 384–388. In: A.M. Prokhorov (ed.). *Physical Encyclopedia in 5 volumes*, Vol. 3. Big Russian Encyclopedia, Moscow.

Laso, M. 2009. *Smart Materials Modelling*. J. Wiley & Sons, NY.

Lazzari, B., M. Fabrizio and A. Morro. 2002. *Mathematical Models and Methods for Smart Materials*. World Sci. Publ., New Jersey.

Marrows, C.H., L.C. Chapon and S. Langridge. 2009. Spintronics and functional materials. Materials today 7/8: 70-77.

Mezzane, D. and I.A. Luk'yanchuk. 2008. *Smart Materials for Energy, Communications and Security*. Springer, Berlin.

Miguilin, V.V. 1998. "Electromagnetic waves". In: A.M. Prokhorov (ed.). *Physical Encyclopedia in 5 volumes*, Vol. 5. Big Russian Encyclopedia, Moscow.

Motyl, E. 2000. Charge in solid dielectrics analysis of measurement methods and investigations. Dr. Sci. Thesis. Wroclaw Polytekhnika, Poland.

Paton, B.E. and A.Ja. Nedoseka. 1994. Main trends of studies of methods and means of technological diagnostics when estimating state of structures and devices. Technological diagnostics and nondestructive control 1: 5-22.

Pinchuk, L.S. and A.S. Neverov. 1993. Polymeric Films Containing Corrosion Inhibitors. Chemistry, Moscow.

Pinchuk, L.S., V.A. Struk, N.K. Myshkin and A.I. Sviridenok. 1989. *Materials Science and Structural Materials*. Higher school, Minsk.

Prokhorenko, P.P. (ed.). 2003. Achievements of Physics of Nondestructive Control. Technolgy, Minsk.

Sacellarion, A., Ch.H. Arns and A.P. Sheppard. 2007. Developing a virtual materials laboratory. Materials today 12: 44-51.

Shil'ko, S. 2011. Adaptive composite materials: bionics principles, abnormal elasticity, moving interfaces. pp. 497-526. In: P. Teešinova (ed.) *Advances in Composite Materials. Analysis of Natural and Man-Made Materials*. InTech, Rijeka.

Singh, J. 2005. *Smart Electronic Materials: Fundamentals and Applications*. Univ. Press, Cambridge.

Skulachev, V.N. 1998. Bionergetics. p. 561. In: I.L. Knunianc (ed.). *Chemical Encyclopedia in 5 volumes*, Vol. 1. Soviet Encyclopedia, Moscow.

Trofimova, T.I. 1990. Course of Physics: University Manual. Higher school, Moscow.

Vanshtein, B.K. 1990. Crystalline structure. pp. 503–506. *In*: A.M. Prokhorov (ed.). *Physical Encyclopedia in 5 volumes*, Vol. 2. Big Russian Encyclopedia, Moscow.

Varadan, V.K., K.J. Vinoy and S. Gopalakrishnan. 2006. *Smart Material Systems and MEMs: Design and Development Methodologies*. J. Wiley & Sons, NY.

Wang, Z.L. 1998. *Functional and Smart materials: Structural Evolution and Structure Analysis*. Kluwer Academic Publishers, Boston.

Wiener, N. 1948. *Cybernetics or Control and Communication in the Animal and the Machine*. MIT Press, Cambridge

Zubarev, D.N. 1998. Thermodynamics of non-equilibrium processes. pp. 87-89. In: A.M. Prokhorov (ed.). *Physical Encyclopedia in 5 volumes*, Vol. 5. Big Russian Encyclopedia, Moscow.

chapter two

Main classes of smart materials and engineering systems

The spheres of materials application are the main attributes of their classification. The seven paragraphs of this chapter each dealing with one of the main (by application) classes of engineering materials, demonstrate the examples in the relevant light squares of Table 1.1.

2.1. Structural materials

Structural materials possess a set of mechanical properties each assuring the performance and the specified life of the pieces bearing mechanical loading. They should comply with interrelated technological and economic requirements. Their aim is to assure minimal energy consumption during the manufacturing process and the necessary quality of individual pieces. The economic requirements relate to the cost and availability of materials because it is essential for mass production of items.

2.1.1. Self-healing materials

This class of SM has been developing intensively in recent decades because they are capable of healing their structural damages caused by mechanical loading. The examples of such materials are given below.

A_1B_1 are the SM, the self-healing of which is initiated by feedback system (FBS) when the stresses exceed the critical value.

The first smart materials of this class were developed on the basis of high-strength reinforced plastics the use of which has reached 15 % of the total in the aerospace industry (Marsh 2002). Their matrix is filled with microcapsules of two types: some of them contain monomers, others contain polymerization catalysts (Fig 2.1). When the stress in the material exceeds the matrix strength limit, micro cracks appear (the primary restructuring). The cracks grow and microcapsules break. The monomer and the catalyst penetrate into the crack driven by capillary pressure, mix up and produce the healing reagent. It adheres well to the polymeric matrix, cures by polymerization and joins the crack walls stopping its growth (Could 2003). The material gets stronger the more the number of healed cracks (the secondary restructuring). The feedback function is fulfilled by the process of destruction of microcapsules and appearance of the healing reagent. The article in this way restores up to 90 % of its original strength (Jacob 2002). Techniques of computer analysis of SM have been developed; they are provided with strength control function and capability to self-heal damages (Balazs 2007). This idea was completed in the 20th century by experimental validation and industrial testing. At present, the strategy of such SM developing, the ways of controlling their properties and main areas of their application in technology has been outlined (Ghosh 2008).

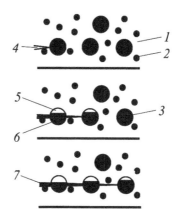

Fig 2.1. Self-hardening polymeric composite (Could 2003): *a* – micro crack origination, *b* – crack destroying microcapsules, *c* – binding microcrack walls with healing agent. *1* – polymeric matrix, *2* – catalyst, *3* – monomer, *4* – crack, *5* – vacuum, *6* – healing agent, *7* – cured healing agent has bound the crack walls

A_1B_5 – the SM which internal friction parameters are expediently changing when loading to ensure the hardening of materials.

The internal friction in solids is the property to convert heat into mechanical energy irreversibly transferred during deformation. The latter dictates the primary restructuring and upsets the thermodynamic equilibrium in the solid body. The relaxation process occurs naturally tending to restore the solid state to equilibrium (the feedback of deformation nature). The dislocation theory of internal friction assumes that its source is the dislocation deceleration (Nowick and Berry 1972). The occurrence in the material phase inhibiting the shift of dislocations (the secondary restructuring) augments the internal friction. Let us show examples of the SM hardening at loading.

The materials subjected to cold-hardening contain the feedback responding to plastic deformation by expanding the limits of proportionally and yielding, augmenting the surface layer hardness bearing the loading (therefore, these SM can also be referred to the taxon A_1B_7). The feedback initiates the multiplication of dislocations making their shifting difficult. The deformation resistance $\sigma = G\sqrt{\eta}$, where G is shear modulus, η – density of dislocations growing together with deformation (Belyaev 1988). Cold-hardening is implemented in smart processes of metals treating under pressure, such as drawing, dragging, rolling, etc.

The hardening plastics serve to produce articles subjected to fatigue sign-variable stresses, such as ship masts, jumping poles, resilient boards, damping machinery pieces, etc. The feedback optimizes relaxation processes in the smart plastic managing the shifting of dislocations. Smart distribution of their concentration ensures the assigned parameters of elasticity within the scheduled range of deformation and peak deformations (Moskvitin 1981).

2.1.2. Heat-resistant materials

Heat resistance is the stability or insignificant impairment of deformation and strength indices of materials under effect of high temperatures ($T > 0.3\ T_{melt}$). The mechanical loading at high temperatures produces relaxation and creep. The temperature growth reduces the stage of creep acceleration finishing in destruction. The long-term strength is characterized by the time the specimen is loaded until it destroys at the fixed stress strain state and at given temperature.

The mechanism of materials high-temperature deformation at long-term loading is due to the motion of dislocations. The sliding of dislocations in the creep plane is decelerated by the dislocation intersecting this plane. The movement of those and other dislocations is activated thermally. The steady rate of high-temperature creep is (Rozenberg 1994):

$$\dot{\varepsilon} = A \exp\left(-\frac{\Delta H}{kT}\right)\sigma^n,$$

where ΔH is the creep activation energy, $T > 0.5\ T_{melt}$ – temperature, k – Boltzmann constant, σ - stress, T and n – constants independent on σ, $n = 4 \div 5$ for metals. The so-called diffusive creep begins at the temperature close to the melting one and the stresses $\sigma\ /\ T < 10^{-6}$. It is described by the equation above with $n = 1$. The creep evolves without participation of dislocations due to the diffusive transfer of atoms in the field of stresses gradient, and determines the change in specimen shape.

It is apparent that the structure correction of heat resistant SM should be carried out by the feedback initiating formation of particles of a new phase decelerating the movement of dislocations and stable at high temperatures.

A_3B_2 are the SM which under high-temperature long-term loading, cause actuation of the feedback optimizing the material structure based on the criterion of resistance to creep.

Heat-resistant steels and alloys contain dispersed carbide and intermetallic phases which appear with participation of alloying elements. These phases inhibit the creep dislocation mechanism augmenting the energy of bonds between particles of solid solutions. The hardening effect in steels and alloys on the nickel base results from formation of secondary carbides (Me_2C_6, Me_6C, Me_7C_3) and intermetallic phases (Ni_3Ti, Ni_3Al, Ni_3Nb). They apply the brake to the movement of dislocations. It can be believed as a result of action of the feedback triggered at critical temperature and deformation values.

The pieces under load operating at $T = 450 \div 470 \ °C$ are made from high-chrome steels. The alloying elements (V, W, Mo, Nb, Ti) are introduced yielding the Laves phase (the feedback) promoting steel heat-resistance.

Superalloys (nickel, cobalt, iron and nickel) contain up to $10 \div 12$ alloying elements and withstand temperatures $1000 \div 1100 \ °C$ and higher. The alloying elements produce with nickel the main hardening γ'-phase. The smart response of superalloys to the effect of high temperatures is that the γ'-phase strength augments as the temperature rises. Application of the super alloys has enabled to prolong the life of combat aircraft jet engines to ten thousand hours. Owing to these SM, reliable operation of turbopumps in the liquid fuel missiles and in the aircraft launching power unit are ensured at the present time (Struk et al 2010).

The prospects of development of heat-resistant metallic SM are related with the provision of the feedback which, when creep appears, would initiate the displacement of dislocations normal to the material sliding planes.

A_3B_{10} are the SM which, when heated, would actuate the FBS optimizing the material structure on the basis of strength criterion.

Materials with elevated melting temperature are needed in nuclear power engineering, aircraft and space engineering. The temperature above iron T_{melt} have Ti, Zr, Hf, V, Nb, Ta, Cr, Mo, Re, Os, Rh, and metals of the platinum group. The article from them are operable at $T > 1000 \div 1500 \ °C$, they are characterized by low heat-resistance. It is due to the fact that at $T \approx 400 \div 600 \ °C$ they undergo intensive oxidation. Chrome with the heat-resistance up to $1000 \ °C$ corresponds to the criterion of SM due to the formation of high melting consistent oxide film Cr_2O_3 on the article surface. The development of these materials presumes creation of feedback system promoting the surface energy and directing it at article passivation.

A_3B_{12} are the SM the heat-resistance of which is adjusted by the FBS checking the process of molten material solidification.

The amorphous metals are solid non-crystalline metals and their alloys. They are thermodynamically metastable systems. They exist due to the retarded processes of crystallization.

Metallic glasses are of most interest as the constructional heat-resistant materials. They include, in the first place, the systems $M_{1-x}Y_x$, where M is transition or noble metal, Y is non-metal, $x \approx 0.2$, and, secondly, the alloys of transition metals Ti–Ni, Zr–Cu in some composition intervals. The yield strength and breaking points of metallic glasses are very high and close to the theoretical strength of base alloys. It is due to this that the glasses have no crystal boundaries and dislocation. The crystallization time of metallic glasses at 300 °C is rated in hundreds years [Gunther and Beck 1983).

The metallic glasses for construction purpose are obtained by pressing of amorphous particles forming by quickly cooling molten drops. The drops appear by splashing melt streams by quickly rotating cold substrate. Fig 2.2 shows the diagrams of spinning (*a*) when the cold substrate is the drum and extraction (*b*) when melt drops are being thrown away by the rotating disk.

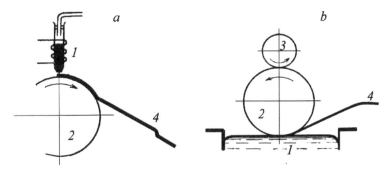

Fig 2.2. Techniques of producing metallic glasses: *a* – spinning, *b* – extraction: *1* – melt, *2* – rotating disk, *3* – auxiliary cleaning disk 2, *4* – amorphous metal

Strictly speaking, the metallic glasses should not be referred to the category of SM. 'Smart' is their production process in which the feedback function is performed by the systems adjusting the melt dispersion and the drops solidification rate.

A_3B_8 are the SM which heat-resistance is controlled by the FBS optimizing the material internal friction.

The heat-resistant industrial ceramics are based on oxides and oxygen free metallic compounds (carbides, nitrides, borides and cilicides). They contain crystalline and glassy phases as well as gas inclusions. The melting temperature of the crystalline phase of oxide ceramics is within 2000÷3300 °C, the working temperature is 0.8÷0.9 of the melting temperature. The main drawback of the ceramics is brittleness. When using original powders very finely milled and optimum baking conditions, it is possible to produce ceramics with the crystals of the size of about 0.3 μm characterized by the plasticity (Yanagida 1986). The prospects of improvement of ceramics performance relate to the application of nanotechnologies, in other words, to produce the structure consisting of nanodimensional (10^{-9} m) particles. The production of nanoceramics is a 'smart' production process in which the functions of feedback system are fulfilled by the dispersion of nanoparticles in the process liquid and their maximum compaction when producing and baking workpieces.

2.1.3. Cold-resistant materials

The equipment performance at negative temperature is determined largely by the cold brittleness of materials, or by brittleness growth as the temperature reduces. At first this problem attracted notice at the end of 19th century due to intensive railway construction and a large number of failures in the wintertime due to the destruction of rails, bridges, tanks storing petroleum product, etc. It still remains the problem today for gas and oil production, aviation and equipment operating in the polar region where the air temperature can drop to −60 °C. The bodies of space missions cool down the liquid oxygen temperature (−183 °C), the units of equipment separating liquefied gases – to the liquid helium temperature (−269 °C). In these conditions the materials lose plasticity and viscosity, their yield strength intensifies and they tend to brittle failure. The interatomic distances in the crystalline lattice reduce; the interactions which are masked at usual temperature intensify by the thermal motion of atoms (Solntsev et al 1999).

The dynamic theory of crystalline lattice explains the cold brittleness according to which the Debye temperature θ_D of the matter physical parameterю Шt, is determined by the ratio $k\theta = \hbar\omega_D$, where $\omega_D = \bar{u}\,(6\pi n)^{1/3}$ is the limiting frequency of elastic oscillations of crystalline lattice, n – number of atoms in the volume unity, $\bar{u}$ – the averaged speed of sound in the solid body, k and $\hbar$ – Boltzmann and Plank constants. The Debye temperature divides the high temperature region $T \gg \theta_D$, where the classic statistical mechanics is justified and the solid plastic heat capacity is described by the Dulong-Petit law, from the low temperature region $T \ll \theta_D$. The quantum effects, quantum statistics and Debye heat capacity laws are effective in this region (Goldade and Pinchuk 2009). Above θ_D, all the modes (types of lattice oscillations) are excited in the crystal, below it, some modes begin to 'freeze out'. Consequently the plastic fracture region of the materials decreases and their yield limit approaches the strength limit provoking brittleness. At the cryogenic temperatures (lower than 120 K) the constructional materials can be treated as atomic and molecular structures with the strength influenced considerably by the 'gas' of elementary excitations (quasiparticles).

The cause of steel cold brittleness from the viewpoint of materials science is the diffusive mobility of sulfur and phosphorus impurities which compounds adsorbed over grain boundaries act as concentrators of stresses. At low temperatures, the local stresses over the concentrators augment drastically weakening the steel cohesive strength (Solntsev et al 1999). To avoid it, it is desirable in operation to initiate in the steel structure the following processes (Struk et al 2010):

- grain refinement or, at least, reducing the steel tendency to growth of grains;
- formation of carbide and carbonitride particles over the grain boundaries to prevent growth of grains;
- nitrogen release from solid solutions and its transfer to the nitride structure, etc.

These processes evolve when producing the materials and unfortunately, they are impossible in the constructional material, the more so because the lowered mobility of atoms. Therefore, one of the few mechanisms of controlling the cold-resistance is to adjust the structure of materials with the quantum mechanics techniques. The feedback system should be actuated in SM 'defreezing' the crystalline lattice oscillations in order to intensify the diffusion of impure compounds deep into the grains and thus broaden the range of stresses and deformations corresponding to viscous fracture of the material.

One more opportunity of creating the feedback system adjusting the brittleness of materials relates to the polymorphous transformations, in other words, to changes in the atomic crystalline structure due to heat absorption or generation. The material structural transformation into a more stable modification is connected with overcoming the energy barrier at the expense of thermal fluctuations. When this opportunity is small, the non-equilibrium phase can exist for a long time in the metastable state. For instance, the diamond appearing at $T > 1200$ °C and $p \approx 100$ MPa, exists for unlimited time at the room temperature and atmospheric pressure without inversion into stable graphite under these conditions.

The instances estimating the development of cold-resistant SM or smart engineering systems operating at low temperatures are discussed below.

A_2B_{10} are the SM in which temperature changes actuate the feedback system optimizing the material structure.

The tin pest is the polymorphous transformation of the tin tetragonal lattice into the cubic one at $T < -12$ °C. This transformation evolves with the maximum rate at $T < -33$ °C and is accompanied by abrupt tin strength change when it crumbles into the gray powder. This phenomenon caused the tragic death of the British Antarctic researcher R.F. Scott and his colleagues in 1912. On the way back from the South Pole they froze because the vessels with gasoline sealed with tin and left on the route, turned out empty. The feedback system should be actuated in the smart tin adjusting the nature of chemical bonds between atoms; in other words, the interaction of their external valence electrons. At the crystallochemistry state-of-the-art this feedback system does not seem feasible to develop.

A similar situation exists in the domain of smart *cold-resistant plastics*. Most plastics for constructional purposes transform into the glassy state at low temperatures and become brittle. The exception is teflon $[-CF_2-CF_2-]_n$, the cold-resistance of which is due to the strong chemical bonds in the carbon chain and between the atoms C and F. To develop a smart cold-resistant plastic it means to interfere with the feedback system in the nature of interatomic bonds, which in fact is transformation of the original polymeric material into the other in operation.

Summarizing, it is noteworthy that the smart cold-resistant materials for constructional purposes have not yet been developed but the concept based on adjusting of atom-electron interactions in the crystalline lattice of overcooled materials is elaborated using quantum mechanics techniques.

2.1.4. Radiation-resistant materials

The resistance of engineering materials to radiation is their capacity to preserve performance properties (primarily mechanical and electrical, optical, chemical and other properties for multifunctional materials) when exposed to the emission of high energy (X-rays, γ-rays, α- and β-particles, accelerated electron, protons, etc.). The modification of properties is due to the displacement of atoms in the crystalline lattice (the radiation-induced defects), ruptures of chemical bonds, nuclear reactions, etc.

The electromagnetic emission (optical photons, γ- and X-ray quanta) first excite the crystal electron system, afterwards the mechanisms actuate atoms displacement in the crystalline lattice: atom interaction with the high energy electron, ionized atom displacement due to electrostatic repulsion from the similarly charged ions, shifts of the simultaneously ionized neighboring atoms, etc.

The materials are very strongly damaged by the neutron and γ-irradiation. The extent of damage depends on the neutron fluence (the number of neutrons hitting per time unity the surface unity of the specimen) or the absorbed dose of γ-radiation.

The appearance of single defects in the crystalline lattice causes the hardening and reduces plasticity of metals. The electric resistance of fine metals augments due to radiation induced defects. The concentration of point defects in semiconductors increases under the radiation changing the electric and optic properties (Kostyukov et al 1979). The inorganic glasses lose transparence due to radiation, acquire coloring and crystallize. The silicates experience anisotropic expansion and solid-phase amorphization reducing the density, elasticity, and heat conductivity. Similar, but less intensive changes are typical for oxides. Concretes are resistant to radiation when exposed to the neutrons with the fluence less than $3 \cdot 10^{19}$ cm^{-2} (Sychev 1994).

The radiation excites and ionizes molecules of organic materials accompanied by gas emission. The resistance to radiation of engineering polymers depends on the concentration of dissolved oxygen and chemical-radiation oxidation of macromolecules. The most significant changes in the structure of plastics occur due to the irreversible processes of bonding and destruction of macromolecules. The reversible changes depend on the power and the dose of radiation and correspond to the equilibrium between formation and decay of unstable products of radiolysis (Pleskachevskii et al 1991). The post radiation ageing of plastics is induced by chemical reactions of free radicals forming when macromolecules interact with air oxygen. The polymeric dielectrics become brittle and lose strength before the electric properties change considerably.

The feedback system in the SM resistant to radiation should respond to any radiation damage by initiating the physical and chemical processes dictating the self-healing of the radiation defects.

A_4B_{13} are the SM which, when radiated, have the FBS initiating the restructuring compensating the radiation damage.

The materials of nuclear power plants, particularly those in the reactor zone, are exposed to the cooling corrosion active media (water at T_{boil}, liquid metals, salt solutions), mechanical stresses, vibration, powerful stream of high energy neutrons, temperature gradients (Todd et al 2010). A typical damage in these conditions is that metallic pieces grow in volume. This effect called the 'radiation swelling' is the result of rupture of metallic bonds and displacement of atoms and appearance of primary radiation defects of the Frenkel couples type: the vacancy + interstitial atom. The interstitial atoms interact with the appearing dislocations, while the vacancies combine producing microspores. The accumulation of primary defects accelerates phase transitions in metallic alloys rendering the brittle and inducing pieces fractures.

The Los Alamos National laboratory (USA) has developed the computer model of SM capable to self-heal the radiation defects (Bradley 2010). The base of the SM is a copper nanocrystalline alloy. The self-healing mechanism is the actuation of intergranular boundaries occupying a vast area. The recombination mechanism of defects accumulation and their absorption by boundaries moving during material exposure is energy-wise more favorable than the annihilation vacancies on diffusion mechanism. The development of such SM is in the stage of laboratory experimentation.

The fuel element (FE) is the main component of nuclear reactor generating heat by nuclear fission of the nuclear fuel. The most used FE are rod shaped (Fig 2.3), passing through the reactor active zone. The hermetic shell renders the FE mechanically strong and protects the fuel from contact with heat carrier dissipating the thermal energy. The shell material (zirconium alloys, stainless steel) has a low neutron capture cross-section, and within the working temperature range, does not react chemically with the fuel or the heat carrier (Kazachkovskii 1998). The appearance in the shell of radiation corrosive cracking results in the neutron leaks, radiation contamination of heat carriers and reduces the reactor efficiency (Todd et al 2010).

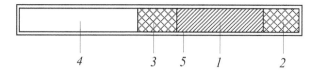

Fig 2.3. The FE of fast reactor (Kazachkovskii 1998): *1* – nuclear fuel, *2* and *3* – screens (depleted uranium),*4* – gas collector, *5* – shell.

It is a topical problem to develop SM for the FE shells. It should be provided with the feedback system 'eliminating' the neutron catching centers from the material structure as the radiation ageing advances.

The radiation absorbing materials (RAM), when interacting with electromagnetic waves of the microwave frequency band, absorb the electromagnetic energy which is partly transformed into the thermal energy. A powerful radiation can lead to heating of the RAM and to thermal deformation of the electromagnetic screens made from these materials (Alekseev et al 1998). The smart RAM should react by initiating the feedback system which actuates the restructuring of materials, for instance, by initiating the radiation-induced chemical reactions (A_4B_{15}) causing the growth of RAM conductivity and reducing the Joule loss.

2.1.5. Corrosion-resistant materials

The corrosion-resistant materials possess stronger chemical resistance, that is why they are able to operate in the corrosion active environments without any extra protection. Let us characterize in brief the physical and chemical mechanisms of damage by corrosion of engineering materials.

Corrosion-resistant materials are chrome, chrome-nickel, chrome-manganese-nickel, and chrome-manganese steels. Their resistance to the electrochemical corrosion is determined primarily by the properties of passivating surface layers on articles. When the Cl^{-1} ions damage the passivating film, the pitting or slot corrosion appears, and at $T > 80$ °C steels crack (the corrosion under stress).

Copper is resistant to electrochemical corrosion in the air, as well as in fresh and seawater at slowly flowing rate. When the liquid flows with the rate $v > 1$ m/s, the so-called jet corrosion begins. Brass is more stable to jet corrosion and not submitted to zinc loss in the chloride aqueous solution and corrosion cracking in the ammonium medium. Bronze is not subjected to jet and gas corrosion.

The corrosion resistance of aluminum is governed by the properties of its oxide film. It can be destroyed by halogen ions inducing the pitting corrosion at elevated temperature.

Titanium and its alloys are passivated in the oxidizing and neutral liquid media manifesting high resistance in the chloride aqueous solutions up to $T < 120$ °C. The titanium does not tend to corrosive cracking.

Nickel is passivated in oxidizing liquids. It is the main engineering material capable of operating in the media containing fluorine and inorganic fluorides.

The high temperature atmospheric corrosion is accompanied by formation of oxides (heat scale) on the article surface. The oxidant diffuses through the oxide layer towards the article surface, the metal atoms diffuse in the opposite direction; the diffusion evolves faster over the crystallite boundaries.

The gray irons doped with Si and Al are resistant to gas corrosion up to $T \leq 850$ °C [10].

The most resistant to the electrochemical and gas corrosion among engineering materials are inorganic glasses, oxides, graphite, and ceramics. The corrosion resistance of ceramics is determined by the chemical inertness of its base. The oxide, carbide, nitride, silicide and other types of ceramics are chemically resistant in aggressive liquids and thermally stable in the air up to the temperatures 1700÷1800 °C (Balkevich 1984).

The high chemical inertness of polymers is explained by the strong bonds in the main chain of macromolecules of these dielectric materials not involved in the electrochemical reactions. The polymers corrode by the mechanisms of swelling in solvents and destruction (the rupture of chemical bonds in the macromolecules) reducing their molecular mass. The thermal destruction of organic polymers in absence of O_2 evolves by the radical-chain mechanism at $T = 230÷430$ °C. In the air at much lower temperature, the multistage chain reaction of thermally oxidizing destruction evolves faster. The mechanisms of other types of destruction, such as photooxidation, hydrolytic, biological – correspond to their names (Shlyapnikov 1990).

The variety of mechanisms of corrosive damage of structural materials presumes the feasibility of implementing numerous feedback systems optimizing the structure of corrosion-resistant SM. Unfortunately, these feasibilities are implemented very weakly.

A_5B_{29} are the SM the electrochemical corrosion of which is decelerated by the FBS adjusting the electrode reaction.

The electrochemical corrosion of metals evolves due to the electrode polarization of metallic pieces – the shift of electrode potential from the stationary value (corresponding to the equilibrium state of the electrochemical system) to the value which is set when the electric current flows through the system. The polarization makes the detail as anode, and electrons are then transferred from the electrolyte to the metal and the metal cations are transferred into the electrolyte. It proves that the detail corrodes.

It is impossible to develop a smart metal which electrode potential stays constant when the current flows in the electrochemical system because this contradicts the energy conservation law. Still, it is possible to develop a smart electrochemical system, which would automatically maintain the electrode potential of a metallic piece within the range of values at which the piece does not corrode. The feedback function in the systems of metals electrochemical protection fulfills the direct external current source or a protective element from other metal electrically connected with this piece.

The feedback in the cathode protection systems shifts the piece electrode potential towards the negative from its stationary value. As a result, only cathode processes evolve in the piece (for instance, H_2 release), while the anode processes inducing corrosion are transferred to the protective element. Some alloys, for instance those of aluminum, have a narrow range of protective cathode potentials. If the potential shift towards negative exceeds the threshold value, the so-called overprotection is possible or the intensive alkalization of the electrolyte surface layer leading to aluminum decomposition. The smart systems of cathode protection prevent the sea and soil corrosion of metallic constructions.

The anode protection shifts the piece electrode potential towards positive direction until the passivating film appears on the surface. Such systems are used to protect the chemically equipment from corrosion in solutions of acids, alkalies and salts (Makarov 1999).

A_7B_{10} are the SM whose swelling is adjusted by the FBS initiating the material restructuring.

The swelling is the solid volume gain due to absorption of gas or liquid from the environment. The swelling can be considered one of the types of damage by corrosion of structural materials as far as the performance properties of the swollen materials deteriorating considerably.

The swelling of polymers leads to the specimen size change while its original shape is preserved. At constant temperature, the extent of swelling grows in time approaching to the equilibrium value. The amorphous unbounded polymers can swell to 2÷3 times of volume. The solid polymer specimen turns into the viscous flow state in the liquid indefinitely compatible with this polymer, while the specimen from bonded polymer increases the volume an order of magnitude under the same conditions. The crystalline polymers swell, as a rule, at the expense of the amorphous part (Papkov 1981).

The monotonous kinetic swelling dependence of polymers at $T =$ const undergoes drastic changes when the restructuring of the chemical or supramolecular structure of polymers takes place. Therefore, the swelling of polymeric materials can be inhibited by the feedback systems developed, for instance, according to the following principles:

- starting the mechanism of the amorphous phase crystallization (secondary crystallization) in partially crystallized polymer by reducing the intermolecular interaction of molecular chains during diffusion of low molecular substances;
- crossing chemical bonding formation between macromolecules (cross-link) by their activation with diffusing particles.

These SM have been developed and tested in the laboratory but have not been adopted for mass production (Struk et al 2010).

The thermally diffusive modification of chemical composition of surface layer of pieces contacting with high temperature gaseous media or plasma is a particular type of corrosive damage of structural materials. It is typical in aviation engines, gas turbines, nuclear power plant reactors, etc. It is impossible to develop the SM without thermal diffusion but it is possible in principle to create the feedback system which would set up the energy barrier to the diffusing atoms. The height of the barrier should be adjustable depending on the temperature gradient in the detail. The most apparent solution to this problem is to initiate the chemical reaction between the material and the high temperature gaseous medium. The reaction should be localized in the piece surface layer and lead to the formation of barrier coating inhibiting diffusion.

The engineering materials capable of limiting high temperature thermal diffusion have been developed, and they are used in thermochemical and thermonuclear missile engines, vacuum tocomac chambers.

A_4B_{26} are the SM whose destruction is adjusted by the FBS changing the intensity of photochemical reactions in the material.

The photooxidizing reactions evolve in polymeric materials exposed to light. One type of these reactions is photodissociation or the macromolecule disintegration along the main chain into radicals. It is the main mechanism of wasted polymers eliminating at dumps. The macromolecule becomes excited and disintegrates along the weak link when the photon with the wavelength 100 ÷ 1500 nm (corresponds the energy 0.8 ÷ 12.4 eV) is absorbed (Kuzmin 1999).

The smart *photodegradable polymer materials* retain stable physical, chemical and mechanical characteristics during operation, but after the service life expires the accelerated photo-disassociation when exposed to light from natural or artificial sources (Goncharova et al 2006). When the number of photo-disassociated macromolecules reaches critical value, the feedback system is activated accelerating the quantum transition of non-decomposed molecules from the main electron state into one excited. The exposed to photodestruction polymeric material disintegrates into fragments without harming the environment with the effect of atmospheric factors. The biodegradable plastics are the example of the situation when the society benefits from the corrosion of materials.

Certainly, the range of SM for construction purposes is not covered by the quoted examples. Because of the book-limited scope, here are just typical cases reflecting the state-of-the-art of structural materials.

2.2. Triboengineering systems

Triboengineering materials are subjected to external friction in operation. The particular external effect on this type of materials is the following.

First, the performance of triboengineering materials in industrial articles is determined mainly by the properties of thin layer on the friction surface. The shear stresses are localized in this layer, which are determined by the resultant force applied to the piece. The mechanical properties of constructional triboengineering materials in the volume of the article are not so significant as the compliance of its surface layer to the shear stress effect.

Second, these materials operate only in the tribosystems. In the classic variant tribosystem is a friction unit consisting of a pair of mobile contacting solid bodies and the 'third body' located in the dynamic contact zone. It means that the feedback system should not be by all means integrated into friction SM; its functions can be fulfilled by other elements of the tribological system.

Third, the problem of triboengineering materials improvement has acquired global significance. The overhauling of machinery and equipment failed due to wear and tear, requires enormous energy and human resources. In all countries it is a severe social burden. Even a moderate reduction of these outgoing would encourage significantly production, engineering safety and environment protections.

These factors evidence the urgency and social importance of developing SM for triboengineering purposes.

2.2.1 Materials for operation at abnormal temperatures

The SM of this class are intended for operation at high (> 500 °C) or low (cryogenic, < 120 K) temperatures or at significant temperature difference. The effect of abnormal temperatures on friction and wear of SM as well as variation of heat expansion-compaction of pieces on the performance of smart friction units is controlled by the feedback systems of different nature.

A_3B_6 are the SM, in which tribotechnical properties deterioration under the effect of abnormal temperatures is compensated by the FBS initiating self-lubrication.

The high temperature friction occurs in internal combustion engines, metallurgical process equipment, space vehicles, etc. under these conditions Petroleum lubricating oils fail to operate: the lubricity of their adsorption layers is limited by the range of temperatures up to 300 °C (Matveevskii 1978).

The high temperature physical and chemical transformations of solid lubricating coatings on bearings parts ambiguously influence the friction. For instance, cobalt having hexagonal shape of crystalline lattice undergoes transformation into the cubic face-centered one at 417 °C. This is accompanied by abrupt growth of friction coefficient of lubricating cobalt coatings. While the weakening of van-der-waals bonds in the nitride complex structure when heated to 1000 °C improves the lubricity of nitride coatings (Silin 1976).

The low temperature friction, in other words, the friction at cryogenic temperatures, is typical for articles of space missiles, nuclear and electronic technology, where liquefied gases serve as working substances or process media. Similar conditions are observed in Arctic, Antarctic and on the Moon, where the temperatures of soil at nighttime fall to minus $150 \div 180$ °C.

The triboengineering characteristics of crystalline materials are affected significantly by the quantum effects evolving at the temperatures lower than the Debye ones (Arkharov and Kharitonova. 1978). A strong cooling stops the mechanisms of chemosorption so the oxide films preventing seizure do not appear on friction surfaces of metallic parts, but the condensate films deposit, which surface energy grows at cooling (Silin 1976). It intensifies the intermolecular interaction between friction surfaces.

Nowadays one of the best antifriction materials for operation at low temperatures (down to minus 200 °C) is the bearing material DU developed by the British company «Glassier Metal». A bronze powder layer saturated with mixture of lead and Teflon is baked to steel band. The functionality of the material at cryogenic temperatures assures the optimum combination of components, cooling does not damage the antifriction and bearing properties. These characteristics are not self-controlled at temperature variations, so the materials DU do not meet the criteria of smart materials.

The feedback system in the virtual SM operating in friction units at low temperatures should direct the energy of quantum transitions in the atomic particles of material at the reduction of surface energy of friction surface parts setting into contact. The Internet sources report that NASA is developing the SM implementing this idea.

A_3B_4 are the SM whose tribotechnical properties impaired at abnormal temperatures are controlled by the FBS optimally restructuring the layer on the friction surface.

The high temperature oxidation of the surface layer on metallic parts can serve as the base of such restructuring. It should be smart because the oxidation effect is ambiguous at high temperature friction: in some cases the oxide film lubricates, in others it undergoes to intensive wear. For instance, the antifriction layer of oxide PbO at $T = 400 \div 500$ °C oxidizes to Pb_3O_4 and loses its lubricity (Silin 1976).

The oxide films on steel, as a rule, have a multicomponent composition dependent on oxidation temperature: wurtzite FeO, hematite α-Fe_2O_3, spinel γ-Fe_2O_3, magnetite Fe_3O_4. The predominance on friction surface of the oxide Fe_2O_3 or the mixture α-Fe_2O_3+Fe_3O_4 reduces the wear rate of steel articles. The spontaneous adjustment of the composition of oxide films formed at high temperature friction heating, serves the base of improvement of operability of friction baked materials which can be referred to the category of smart. They demonstrate a strong heat resistance and high temperature wear resistance at dry friction (Fedorchenko et al 1976).

The high temperature ($T \approx 1000$ °C) tribochemical interaction in metallic materials is a proven means of friction reducing (Heinicke 1984). When sodium chloride melt is used as lubricant, the iron chloride is generated on the friction surface of steel articles. The iron chloride enters into the tribochemical reaction with iron oxides producing a lubricating layer of eutectic mixture of these substances. The lubrication of steel parts with phosphates melt also forms at elevated temperatures an antifriction eutectic layer on the friction surface in which phosphides dominate. The molybdenum disulfide enters at high temperatures into tribochemical reactions with metals producing the metal sulfide and molybdenum: $MoS_2 + 2Me = 2MeS + Mo$. The resulting molybdenum participates in the tribochemical reaction producing a new polycrystalline lubricating layer Me + Mo at high temperatures.

The restructuring of surface layer in these smart systems is controlled by the feedback system actuated at $W > W_a$, where W is the thermal energy applied to the friction surface, W_a – the activation energy of the tribochemical reaction.

A_3B_{11} are the SM whose antifriction behavior at high temperatures is adjusted by the FBS determining the material partly melting on the friction surface and implementing the liquid friction.

Powder bearing materials consist of the rigid porous matrix the free volume of which is filled up with easily melting component. This material reacts to the heating by melting and swelling of the specific volume of low melting inclusions. The liquid phase releases from the pores to the specimen friction surface. The released liquid volume depends on its viscosity and difference between capillary pressures in the matrix pores and in the clearance in the friction couple. The feedback system optimizes the ingress of liquid phase into the clearance of the mobile contact adjusting the lubrication conditions in response to the load and rate of friction.

This idea has been implemented in the SM for bearing inserts with the porous layer from baked bronze powder (Ministr and Priester 1970), mobile electric contacts (Braunovich et al 2007) and contact seals with adjustable tightness (Pinchuk 1992).

A_3B_8 are the SM operating at temperature drops, the triboengineering characteristics of which are optimized by the FBS adjusting the internal friction in the material.

The ski grease is a typical example of such SM. The grease layer is applied to the ski-sliding surface. When classic skiing with definite combination of the sliding speed and the ski pressure on the snow, the optimally selected grease assures the partial melting of the snow layer contacting with the ski. It turns into water, its internal friction drops and the ski slides easily on the snow. Thus, the external friction of the ski is adjusted by the feedback system based on the ski grease property to melt snow on the snow and reduce its internal friction at definite speed and force sliding conditions. At zero sliding speed and larger load (the phase of pushing away), the layer of grease does not affect the internal snow friction and ensures better adhesion to the snow (it is said, the snow 'holds' the ski).

There are thousands of types of ski grease each intended for application to the snow of definite structure within a narrow range of temperatures. No multipurpose ski grease has been developed so far: it should have a feedback system modifying the thermalphysics and tribotechnical characteristics of layer of grease in response to the loading-speed and power parameters, temperatures, structure and deformability of the snow.

A_3B_3 are the virtual SM, which would automatically compensate the unfavorable thermal deformation of friction parts.

The thermal deformations of articles occurring under the heat from friction or other sources upset the congruence of friction surfaces, changed the shape and size of clearances in movable conjugation. There is a new trend of tribodiagnostcs: it is tribodilatometry studying the regularities of clearance changes in friction pairs due to wear and thermal expansion of articles (Potekha 2000). The perfect solution to this problem would be the application of SM which thermal expansion changes according to the program set by the feedback system. It is unlikely that such materials will be created in the near future.

This problem is resolved when the friction units are designed by justifiable choice of materials and setting of clearances, developing cooling and heat-exchanging systems, etc.

2.2.2   Materials exposed to fields and radiation

The information about the effect of powerful nuclear emission on the performance of friction units is rather scanty (Todd et al 2010, Drozdov et al 2010). Such data are practically absent in scientific publications because such tribosystems are not widespread and the information is restricted. The current flow through mobile electric contacts has a specific influence upon the wear (Braunovich et al 2007). In a number of cases, the tribosystems are subjected to targeted influence of fields and emissions to improve their performance or impart them new properties. Given below are the examples of such types of smart tribosystems.

A_4B_6 are the SM which lubricity is optimized by the FBS operating owing to the energy of external emission.

The phenomenon of abnormally low friction in vacuum was registered in the USSR in 1969 as discovery. It appears in friction couples metal-polyethylene (or molybdenum disulfide, or graphite) operating in vacuum irradiated with a flow of accelerated helium ions, which induces a considerable energy release ($10^4 \div 10^6$ kJ/kg) in the surface layer (the thickness < 100 nm) of the polymeric part. The friction coefficient drops from the value 0.1 typical for these couples to the value 0.001 (Dukhovskoi et al 1969). This phenomenon takes place when friction couples operate in the space. The super low friction is controlled by the FBS, which represents the deceleration of helium ions in the material layer on the friction surface. It localizes large energy at ions decelerating fast. The original value of the friction coefficient restores in $10 \div 100$ seconds after irradiation ends. This phenomenon was used for developing a group of solid smart lubricating materials intended to operate in the radiation conditions.

A_5B_5 are exposed to electric fields smart friction units which triboengineering characteristics are adjusted by the FBS regulating the 'third body' structure.

The selective wear in friction was discovered in the 1960s by D.N. Garkunov and I.V. Kragelskii in the friction couples 'steel – copper alloy' lubricated with the alcohol-glycerin mixture preventing copper oxidation. A copper film ~1 μm thick appears on the steel part surface at the steady friction conditions: the sliding speed up to 6 m/s, pressure $30 \div 40$ MPa and temperature $40 \div 70$ °C. The copper crystalline lattice in this film called 'servovit' contains many vacancies: more than 10 % of latticee points are vacant from atoms. The wear of this couple reduces to the servovit film transfer from one friction surface to another. Due to this the friction coefficient lowers to the values typical for liquid lubrication or it determines the effect of wearlessness.

The smart selective transfer system operates under control of the electric field which has appeared in the galvanic couple "steel – alcoholic mixture – copper". The servovit film consists of copper colloid particles, which are the product of anode reaction in this combination; it forms following the mechanism of electrophoresis in the field of this reaction (Litvinov et al 1979).

The selective transfer is implemented in some friction units of automobiles, aircrafts, ships (reducing gears, pumps, etc.). The new lubricating materials improve the wear resistance of friction units in many cases without modifying their design.

Triboelectrization of dielectrics has a multifactor and controversial effect on friction and wear. The movable contacting electrified parts interact electrostatically, the induced electric field affects considerably the wetting of friction surfaces with lubricating liquids, wear particles traveling in the dynamic contact and appearance of the 'third body' (Mironov and Pleskachevskii 1999).

One instance of the smart system of this class is splitting of diamonds. The diamond crystals are cut off with a rotating thin metallic disk with applied abrasive coating. The following was quite unexpected. When the disk is fixed between two differently charged electret plates, the cutting process accelerates noticeably, and the quality of the diamond fresh surface improves (Bocharov et al 1984). The FBS role is played by the electret charge of the plates which minimizes the disk wear and increases the diamond wear. The tribophysical mechanisms of this phenomenon need further study, though the phenomenon itself of accelerating processing of diamonds in the electric field has been confirmed by experimentation and it is employed at the Kristal Works (Smolensk, Russia) (Mironov and Pleskachevskii 1999).

A_5B_6 are the SM of which availability in electric fields is improved by the FBS adjusting lubrication.

The movable electric contacts – sliding and commuting (switching over) – provide the adjustable electric power supply to radioelectronic and electroengineering devices. The movable contacts should fulfill contradicting requirements of smallest wear and stable electric resistance in the contact. The repetition work of electric contacts needs large amounts of noon-ferrous and noble metals irretrievably lost due to erosive electric wear and tear, therefore satisfaction of these requirements has a large economic importance.

The new generation electric contacts are fabricated from composite materials containing solid lubrications: carbon-graphite, metal-graphite, on the metallic powder matrix saturated with low melting alloys, and others (Braunovich et al 2007). They are being improved and provided with FBS which is actuated by electric field and performs the following:

- adjustable release of conductive lubricating components on the friction surface;
- adjustable tribo-oxidation of solid lubrication;
- stabilization of electric resistance by advisable distributing of particles (products of electric erosion wear) in the electric contact, etc.

The friction couples 'ceramics – ceramics' are characterized by high rigidity of movable interface, stable availability at elevated temperatures, endurance in aggressive media (Jahanmir 1984). The friction of articles from dielectric ceramics is inevitably subjected to triboelectrification. The tribocharge field affects the wetting of friction surfaces considerably by polar lubricating liquids. The electret charge on the smart ceramic couples improves lubrication activating adsorption and intensifying the adhesion of molecules in the lubricating material to the friction surface.

It is particularly essential for joint prostheses with a precision friction pair consisting of precisely conjugated ceramic ball and spherical cavity. Irrespective of the biological liquid occurrence in the joint cavity, the implanted endoprostheses with the ceramic friction pair usually work under the conditions of lubricant deficit (Pinchuk et al 2006). The FBS function in such friction units is fulfilled by the tribocharge initiating the adsorption formation of boundary lubrication layer on the endoprosthesis friction surfaces.

A_6B_5 are the friction SM and smart tribosystems the wear resistance of which is controlled by the FBS with external magnetic fields directionally modifying the clearance in the friction couple.

The magnetic suspension arm is the contactless suspension of mobile part over stationary counterbody. The ferromagnetic moving parts are suspended with clearance over the bearing implementing the principles of repulsion or attraction with the help of constant magnets or electromagnets including those with super conducting coils (Uskov and Bogdanov 1995).

The smart magnetic suspensions are provided with the FBS adjusting repulsion or attraction of magnetic components of mobile piece and support. The signal actuating the FBS is the critical friction value in the movable interface.

A_6B_6 are the SM with the FBS, which adjusts lubrication consuming the energy of external electromagnetic or magnetic field.

The synovial liquid contained in the void of joints of man and mammals lubricates the articulate cartilages in the organism biophysical field.

A surprising feature of the synovial liquid is the restructuring and lubricity improvement under electromagnetic field effect. It is demonstrated *in vitro* with the pendulum friction machine with the incorporated electromagnetic coil on the support. When the machine is switched on, the friction coefficient in the pendulum support lubricated with the synovial liquid reduces in time to some constant value (Fig 2.4). This feature is not typical for low molecular liquids such as the salt solution (curve 1). The size of friction coefficient reduction from initial value depends on the magnetic sensitivity of the synovial liquid. It is determined by the degree of pathological change in the joint from which the synovia was taken (Chernyakova et al 2011).

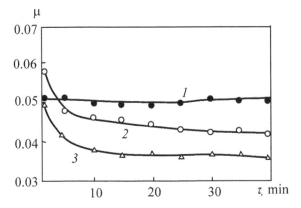

Fig 2.4. Dependence of friction coefficient in the pendulum tribometer support at the time of electromagnetic field effect on the liquid lubricating layer (Pinchuk et al. 2006): *1* – salt solution, *2* – artificial synovia 'Diasinol', *3* – natural synovial liquid

The magnetic powder lubrication technique envisages the mixing of solid lubricating material with the ferromagnetic powder. The mixture is used in friction units operating in vacuum, inert and aggressive media, under radiation (Drozdov 1979). The lubricating material consisting of powders: 80 % of molybdenum disulfide and 20 % of nickel (or cobalt, or lithium ferrite), is delivered to the friction surface of the magnetized part under magnetic field induction $B \geq 0.03$ T. It permits to increase the volume of the lubricating material circulating in the friction unit and to retain the lubricating layer on the friction surface by magnetic attraction. Each ferromagnetic particle (its magnetized part) connected with the lubricating material by molecular forces, acts as the FBS with the attraction force $F = \chi V H(dH/dx)$, where χ and V are the magnetic susceptibility and ferromagnetic particle volume, H and dH/dx – the tension and tension gradient of the magnetic field in the lubricating layer. The magnetic powder technique is the only lubrication technique suitable for friction units operating on the Moon and Mars (Drozdov et al 2010).

A_6B_8 are the SM whose performance is improved in mobile conjugations at the expense of internal friction decrease under FBS effect consuming the energy of electric or magnetic field.

Slip casting is the technique of fabricating workpieces of ceramic articles by preparing the casting slip (aqueous slurry with the water content $25 \div 45$ %) using clay and ceramic powder, pouring the slip into porous (to remove water) molds and drying the molded workpieces which are afterwards converted into ceramic articles by baking (Shatt 1983). One of the main process problems of slip casting is the utmost compaction of ceramic particles in the mold. It is solved by forming the clay slurry between particles acting as the internal friction regulator during molding. The function of interlayers in the best slips is performed by the 'coat' of molecules of surfactants adsorbed on ceramic particles (Fig 2.5). The FBS in the smart slips acts adjusting the 'coat' density, therefore, it adjusts the boundary friction parameters between particles during molding.

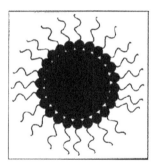

Fig 2.5. Diagram of lubricating layer formed by surfactant molecules adsorbed on ceramic particle

The development of these SM is topical for powder metallurgy because the main problems in this branch relate to the energy cost for powder workpieces pressing and to defects generation in the end articles due to insufficient compaction of particles in the mold (Pimenov 1990).

Extrusion of magnetic plastics, in other words elastic constant magnets on the polymeric binder, is the example of the process system in which, like in the preceding instance, the adjusting function of internal friction is performed by the polymeric melt interlayers between hard-magnetic particles. The layer viscosity in the smart system is adjusted by the FBS (the magnetic field of the particles) assuring the minimal internal friction coefficient at different concentrations of particles in the melt and extrusion process conditions.

2.2.3. Materials for operation in corrosive media

Materials with elevated corrosion resistance operate in corrosion aggressive media without any extreme protective actions from corrosion. In most cases, joint action of corrosion and wear upsets the qusistationary state of such materials.

The corrosive-mechanical wear of metals evolves during friction in electrolytes. This type of wear relates to a large group of essential parts in many spheres of engineering (Garkunov 1985, Lazarev and Preis 1979):

- during sliding friction – end-face seals, pumps of bearings, cylinder and piston units of internal combustion engines, working parts of the processing equipment if the latter contains fatty acids (vegetable oil and animal fat);
- during hydroabrasive wear – stirrers of reactors, wheels and bodies of centrifugal pumps, parts of centrifuges, hydroturbines;
- during gas abrasive wear – parts of air blowers, tilework of chimneys, parts of jet mills and aviation engines, blades of steam turbines;
- during wear by cavitation – ship screws, liners of internal combustion engines, impellers of chemical reactors;
- during fretting corrosion wear – wheel axles and hubs of rolling railway stock, keys and slots, splice joints, riveted sheets, wire ropes, etc.

The mechanism of corrosive-mechanical wear consists in the following. A pair of metallic parts, of which the dynamic contact area submerged in the conductive liquid, represents an electrochemical system; the rubbing articles serve as electrodes and the lubricating layer of the conductive liquid serving as electrolyte. The surfaces of friction articles get covered with coatings of corrosion products, which are damaged mechanically and removed at the wear proceeds. Juvenile (fresh) parts of friction surfaces react chemically with the electrolyte $7 \div 8$ orders of magnitude faster than the ones coated by passivating film (Lazarev 1987). Such electrochemical system is unsteady, the current appears in it and the electrode potential in the friction unit displaces from the stationary by the value ΔU. The current density j serves the measure of the electrode process rate, and the dependence $\Delta U(j)$ is the tribosystem polarization characteristic.

The corrosion-mechanical wear of metals is usually suppressed by the techniques of electrochemistry (Preis and Dzub 1980): by polarizing the tribosystems in order to reduce the energy of friction surfaces, protective elements can used for this purpose; by reducing the area of article (cathode) moving out of contact with the interfaced part; by introduction into the electrolyte of special additives shifting the potential of friction unit into the optimal region, including corrosion inhibitors which can be, by the logic of the instance, called 'wear inhibitors', etc.

These techniques are the base of FBS in smart tribosystems subjected to corrosion.

A_2B_1 are the SM in which mechanical stresses appear at friction 'switching on' the FBS, which reduces the corrosiveness of the medium penetrating into the clearance in friction couple.

Antifriction plastics containing corrosion inhibitors (CI) are the group of SM, which inhibit corrosive-mechanical wear of interfaced metallic counterbody. The surface layer of the article from this plastic is provided with the system of communicating pores felled with the CI. When plastic article contacts dynamically with the counterbody, the micropores release the inhibitor in the liquid or gaseoues phase. The surface active molecules of CI are adsorbed on the counterbody, mainly on juvenile friction spots, compete with corrosive medium molecules penetrating from outside into the friction couple clearance. The monomolecular layer of CI protects the counterbody from corrosion. The CI release from micropores adjusts the FBS actuating when the friction rate and contact stresses on the friction surface reach critical values. The tougher the friction mode and the larger the rate of wear of the passivating oxide layer on the counterbody, the more inhibitor is delivered into the movable contact area (Goldade et al 2005).

A_2B_6 are the SM and smart tribological systems which resistance to the corrosive-mechanical wear is intensified by the FBS neutralizing the corrosive medium aggressiveness by lubricant.

Piston-cylinder-unit of internal combustion engines is subjected to corrosive-mechanical wear when running on the fuel with high sulfur concentration. The sulfur when combusted produces the oxide SO_2 which transforms at catalytic oxidation into the sulfur anhydrate SO_3. When it reacts with vapors of combustion products, it forms the sulfuric acid. The latter condenses on cylinder walls causing its wear near the outlet and inlet ports (Garkunov 1985). It is hard to imagine a metallic material, which would neutralize the acid condensate during friction. But a smart system can be developed to monitor piston-cylinder-unit. It is provided with the FBS, which neutralizes the acid condensed on the vessel wall providing adjustable feeding of alkaline additives into the lubricating oil. The term 'adjustable' implies the amount of the alkaline additive at each moment of time corresponding to the quantity of acidic deposits. To this purpose, the measuring transducers are used connected with alkali dose-meters by means of computer.

A_8B_3 are the SM and smart friction units which are subject to corrosive-mechanical wear eliminated by replacing the external friction of interfaced parts with internal friction of the elastic element?

The stationary joint containing a highly elastic intermediate element ensure the moderate liner and angular displacements of the friction pair in aggressive medium. Fig 2.6 shows the rotary joint in with elastic sleeve *1* seals the space between 'bearing 2 – shaft 3' pair and body 4 isolating the friction areas from the aggressive medium (Garkunov 1985). The angular and linear shaft displacements are fulfilled due to the deformations of sleeve 1. Therefore, the external sliding or rolling friction of bearing 2 against the body 4 which should be accompanied inevitably with corrosive-mechanical wear, is replaced by the internal friction in this sleeve 1. The FBS function in this smart system is fulfilled by the natural dependence of sleeve material elasticity on the deformation value.

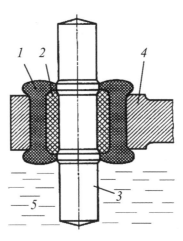

Fig 2.6. Rotary joint with elastic element: *1* – elastic sleeve, *2* – bearing, *3* – shaft, *4* – body, *5* – corrosive medium

2.2.4.   Lubricants

Lubricating materials possess the capability, while being in the fine layer between rubbing parts, to reduce friction and wear, prevent seizure of friction pairs and scoring of friction surfaces. A particular problem solved with lubrication is the friction stabilization permitting to reduce vibration, eliminate sound effects in friction, to assure steady motion of metallic strip when rolling, etc. The perfect lubricating layer should separate friction surfaces, preventing contact between surface patterns. If not, the lubricating layer is destroyed by friction forces, removed (partially or fully) from the friction contact by moving parts, or the properties of the lubricating material are deteriorated by tribodestruction. The tribological restructuring of the smart layer is checked by the control of FBS, which eliminates the damage consequences in the layer using the friction energy.

A_2B_4 are the SM whose corrosive-mechanical wear is controlled by the FBS modifying the rheological parameters of the lubricating layer.

Plastic greases are greasy lubricating materials obtained by the introduction into liquid oils of solid thickeners (soap, paraffin, silica gel, ash, etc.). Their distinctive property is the tixotropy, in other words the capability to restore the original structure destroyed by mechanical effect. The load bearing capability of the lubricating layer of plastic grease is characterized by the following parameters: η_0 – the maximum effective viscosity of the virtually undamaged structure, η_m – the least effective viscosity of the ultimately damaged structure, τ_0 – the limiting shear stress destroying the lubricating layer structure. The dependence on its effective viscosity η on shear stress τ is determined by formula [54]:

$$\eta = \eta_m + (\eta_0 - \eta_m)\,\frac{\tau/\tau_0}{sh(\tau/\tau_0)}$$

At small values τ inducing very slow flow of the lubricating layer, the plastic greases resemble the solid plastic body because the rate of restoration of the original structure exceeds the rate of its destruction. At $\tau \gg \tau_0$ the structure turns out extremely destroyed and the plastic lubricating layer turns into the liquid state with low viscosity η_m. The process of restoration of the destroyed structure at rest is accompanied by the growth of strength growth and the load bearing ability of the lubricating layer. The S-shaped dependence $\eta(\tau)$ within the limits from η to η_0 is described by the fraction in the equation above representing the mathematical model of FBS in the plastic greases.

A_2B_6 are the smart lubricating systems in which the lubricating effect checks the FBS adjusting the entry of lubricating material on the friction surface in response to load.

The human natural joint demonstrates the record low friction coefficient (0.02÷0.05) throughout almost all life span. Its friction couple consists of interfaced microporous antifrictional cartilages the free volume of which is filled up with the lubricating synovial liquid. When the joint is loaded, the synovia is pressed out from the micropores and delivered to the movable contact spot, primarily there where the contact stresses are largest. It is the reverse process when the joint is relieved: the liquid from the clearance between cartilages is sucked into the cartilage microcapillary system (Pinchuk et al. 2006). The FBS function in smart joints executes the property of its synovial medium to establish equilibrium between the capillary pressure in the cartilage micropores and the contact pressure on the friction surface (Chernyakova and Pinchuk 2007).

A_2B_{11} are the SM the parts of which implement the liquid lubrication in the friction units by melting (induced by frictional heating) the layer on the friction surface.

Soft antifriction alloys based on tin, lead, cadmium, etc. (babbits), developed in the 1930s are believed until now the best bearing materials due to their good running-in ability, high heat conductivity and low friction coefficient (Bushe 1993). After reaching the load threshold value, the bearing surface layer from this alloy melts under the effect of friction heat forming a liquid lubricating layer. The same category includes the materials mentioned in 2.2.1 on powder aluminum or copper matrices the free volume of which is filled with the soft antifriction alloy (Pratt 1993). When these SM pass into the mode of liquid lubrication at 180÷200 °C, the viscosity and thickness of the lubricating layer are adjusted by the FBS which indicates the friction heat. However, even being smart, these materials are imperfect. The melt removal from the friction area weakens the shaft fit in the bearing, reducing the precision and rigidity of the friction unit. The next generation of SM or smart friction units will be provided with the FBS compensating the lubricating layer removal.

A_2B_{16} are the SM with the FBS adjusting the lubrication of friction unit by emitting particles and radiation.

The triboluminenscent materials synthesized in 1966 (europium dibensilmetide triethyl ammonium) turned out to be in demand in the 21 century (Fontenot et al 2011). The friction stipulates for deformation and asymmetry of the surface layer crystalline structure in these materials resulting in the ruptures of crystalline lattice and formation of opposite charged carriers. When the lattice is reconstructed, the free charges exits through ruptures to the friction surface producing luminescence. The triboluminenscent lubricating layers are in use as smart sensors adjusting the lubrication effectiveness.

2.2.5. *Frictional materials*

Frictional materials should have a large friction coefficient (usually $f \approx$ 0.2÷0.5) and strong resistance to wear. They are used in braking devices decelerating the sliding velocity of interfaced friction units by transforming the kinetic energy of moving parts into thermal energy and is dissipation. The specific feature of frictional materials is that the critical factor of their availability is the friction coefficient f stability; its drop is impermissible when the sliding velocity and temperature augment. The frictional materials operate at cyclically heating-cooling conditions initiating thermal stresses, which exceed considerably the material mechanical stresses and cause thermal fatigue of materials (reduction of thermal resistance). The stable f values are achieved by reducing the frictional interface contact rigidity. It reduces its inertia measured by the time of growing of the friction moment after the load is applied or moment decrease – after the load is removed. The time of deceleration, the sliding velocity and power changes are the initial data for calculating temperature in the braking device with the equations of frictional thermal dynamics proposed by A.V. Chichinadze (Chichinadze and Braun 1979).

The main types of frictional materials in modern machine building are some kinds of plastics, cast irons and metal-ceramics.

The frictional plastics were manufactured in the 1980s using asbestos fillers, which were later replaced with slag and mineral wool. Caoutchouc, thermosetting resins and their combinations are used as polymeric binding plastics. The binders are filled up with mineral powders (ferrous red paint, chrome oxide, diatomite, etc.) reinforced with wire, fibrous mass, cut cardboard and tissues. The brake blocks are usually produced by saturating the reinforcing mass with catch-up latex or phenol-formaldehyde resin followed by hot curing in molds.

Frictional plastics when rubbing with metallic pairs obtain from 2 to 20 % of generated heat in dependence on the stationarity of friction regime (Chichinadze 1980). Under oxygen effect and heating, the plastic binders undergo the thermal-oxidizing destruction. The products of macromolecules decomposition produce the lubricating layer on the friction surface reducing the f and augmenting the braking device wear.

Frictional cast irons are gray in which a considerable portion of carbon is in the state of graphite, as well as more strong and thermally stable alloyed manganese-phosphor and nickel-molybdenum cast irons. In their operation incorporated into the braking device, there is a risk of cracking of cast iron parts due to the cyclic heating-cooling when the cast iron manifests the tendency to structural transformations (ferrite content growth) at $T > 400$ °C, and accelerated wear of the counterbody takes place under load $p > 0.6$ MPa. Increment of ferrite concentration in the cast iron structure reduces the friction coefficient and raises the probability of brake shoe seizure with the steel counterbody.

Metal ceramics are baked friction materials produced by the techniques of powder metallurgy from dispersed solid particles of metals and non-metals. The typical components of metal ceramics are: the metallic base (iron, copper, bronze), frictional additive to prevent seizure (oxides, borides, and crushed minerals) and solid lubricants. The best materials of this type excel other friction materials because they are able to add the components with various functions ensuring availability of metal ceramics in the composition of the braking device.

The antilock brake system (ABS), well known to motorists, is an instance of smart friction system. It serves to prevent the loss of car control during hard braking. The modern braking systems of vehicles are provided with the hydraulic track and vacuum booster, which permit, at a light pressing of the brake pedal, transmit a large force to the brake shoe blocking wheel rotation. That can provoke the tires to start sliding on paving instead rolling (the so-called skid). As a result the braking time and track extend and the car loses handleability.

The ABS makes the braking more efficient cyclically adjusting the pressure of brake shoes on the wheel disks. The braking cycle of each wheel comprises three phases. At the first phase, the wheel angular speed sensor (perform the sensor functions) sends electric signals to the analytic circuit (actuator) comparing speeds of wheels. If the sliding of at least one wheel is probable, the control unit (processor) stops the feeding of braking liquid into the braking cylinder. As a result, the pressure in the cylinder does not increase even if the brake pedal is pressed.

The second phase is that the control circuit opens the exhaust valve of the brake cylinder. The pressure is released and the wheel rotates faster. If necessary, the FBS is actuated – the braking liquid is pumped from the cylinder into the dumping chamber. As a result the pressure in the braking line drops still faster.

At the third cycle, the control circuit closes the exhaust valve and opens the inlet valve increasing the pressure in the brake cylinder and the pressure force of the braking shoe on the wheel disk.

Such cycle repeats until the braking is completed or the brake shoes stop the locking. The actuator sets the phases duration. The result is a shorter braking path, better maneuverability on the slippery road, improved handleability in case of urgent braking, and ensured tire wear reduction.

In conclusion it should be noted that triboengineering systems are most advanced technical objects from the viewpoint of artificial intellect implementation in engineering. It is conditioned not only by the opportunity of FBS setting within the volume of friction material, but also to incorporate the FBS functions into the tribological system. It is essential that tribology has been created and developed by outstanding individuals beginning with Leonardo da Vinci, I. Newton, M.V. Lomonosov, S. Coulomb, N.E. Zhukovskii, ending with P. Jost, I.V. Kragelskii, A.S. Akhmatov, P.A. Rehbinder, F. Bowden. Just the enumeration of researchers' names whose fundamental works have determined the tribology state-of-the-art would take several pages. The previously mentioned following inventions were registered in the USSR: the selective transfer in friction, abnormally low wear in vacuum. The list can be continued by such outstanding developments in tribology like the theory of fatigue wear, the phenomenon of hydrogen wear and adaptability of friction materials, the elastic-dynamic theory of lubrication, the technology of surface engineering, etc. It has determined the present standard of the science of frictional materials and appearance of its modern tendency of transition to smart tribological systems.

2.3.   Protective systems

The term 'protective systems' incorporates the group of materials intended to protect technical articles from aggressive technological media and environments in the following ways: by preventing contact in the aggressive medium with articles (sealing materials, protective coatings); removal of aggressive components from the medium (filtering and radiation absorbing materials); reduction of aggressiveness of the medium (materials insulating sound and heat, containing corrosion inhibitors).

The usual protective materials are inert to the articles protected and do not affect their operating ability. The SM are capable of modifying the service properties purposefully. The promising trend is the group of SM and smart engineering systems, which function both as shield preventing damage by environment, and interact actively with the environment and reduce its aggressiveness in one or another manner.

2.3.1   *Sealing materials*

Sealing materials prevent the mass exchange between the sealed object and the environment. The sealing element (seal) files up the pores in the joints of articles. It is exposed to intensive physical and chemical action of both the separated media and should conserve its properties unchanged as long as possible. The seal in the interface of mobile parts 'watch' over the surface microroughnesses on the mobile part and forms the antifrictional couple displaying the wear resistance. As a barrier element at the interface, the seal could be affected by all types of external exposures.

Liquid and gaseous sealing SM

Liquids and gases are hard to employ in sealing systems because they are incapable of preserving any shape. With respect to gases, this problem is overcome in the ejecting seals in which the sealed medium is directed away from the joint clearance with the help of the auxiliary gas flow. Gas locks contain the inert gas (helium, neon, argon, krypton) preventing the outflow, mixing, chemical reactions and other consequences of contacting, unlike sealing media in seals.

The liquids find a broader application in the sealing systems other than gases. Their inherent flowability is inhibited by phase or structural transformations when the changes occur in the temperature (melting/solidification), tension of electric and magnetic fields, chemical conversions (polymerization/bonding) or mechanical effects (tyxotropy). Liquids can act as a sealing material without losing flowability. They serve as separating medium when the pressure in front of the seal is lower than the atmospheric or when cutting off hazardous substances (aggressive, explosive, toxic, abrasive) (Pinchuk 1992).

The instances of smart sealing systems based on liquids and gases are discussed below.

A_1B_{1-5} is the class of SM subjected to deformation in seals when the FBS is actuated after the stresses in the seal exceed some critical value modifying in a certain way the bonding strength in the material, therefore, changing its structure, rheological and tribotechnical characteristics.

Thixotropic dispersed systems are diluted irreversibly at intensive mechanical effects (mixing, stirring) and lose fluidity or solidify if left at rest. Thixotropy is a typical property of polymeric and dispersed coagulating structures, which can be destroyed innumerable number of times in the isothermal conditions, and later their properties are fully restore.

The dilatating colloids and dispersed systems with the effect of rheopexy are used as seals which augment rigidity when the pressure drop rises in the seal. Dilatancy is the growth of viscosity of concentrated dispersed systems (pastes) when the deformation accelerates. Rheopexy is the phenomenon of growth strength of dispersed systems subjected to insignificant pressure and slow deformation velocity. The mechanism of the FBS operating in smart dilatating and rheopexic sealing systems consists in the change of density and rigidity of the framework consisting of solid particles distributed in the dispersed liquid.

The liquid rubber is the elastomeric aqueous emulsion based on modified petroleum derivatives with polymeric substances. The Canadian company Lafarge manufactures it (Liquid Rubber^{TM}) to seal joints in building structures. The liquid composition is sprayed into the clearance between blocks by 'cold' technique where it forms a membrane impermeable to water and acid rains, and having a strong adhesion to constructional materials. The smart membrane restores 95% shape after stretching by 1350 % (Boskolo 2006) due to the FBS initiating abnormally intensive rheological processes in the membrane.

A_1B_{11} are the SM of moveable seals of which the degree of tightness is adjusted by the FBS channeling the frictional heat at partly melting the seal and filling the clearances in the moveable joint by liquid phase.

The seal of conjugated conical saddle and rod are used in vacuum systems. The rod is provided on the contact surface with insert of low melting metal which melts at friction. Wood alloy, tin, silver, indium and other low melting metals and alloy are used as inserts (Golubev and Kondakova 1986). Their melts fill up the clearance in the conical joint rendering its vacuum tight.

To add artificial intellectual properties to these adaptive sealing systems, it should be provided with the FBS, which would adjust the melting process of inserts in response to the vacuum degree in the sealed volume.

$A_{5-6}B_2$ are the liquid seals in which the degree of tightness is adjusted by the FBS optimizing the liquid interlayer position in the sealing clearance with the help of electromagnetic force.

The liquid-metallic seals developed in the 1970s (Orlov 1977) are broadly used in aviation and space machine building. Fig 2.7 shows one of the seal varieties (Belyi et al 1980).

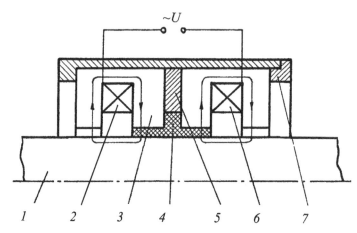

Fig 2.7. Induction liquid-metallic seal: *1* – ferromagnetic shaft; *2, 6* – excitation coils of electromagnets; *3* – magnetic conductor; *4* – T-shaped porous ring with liquid metallic filler; *5* – diamagnetic separating sleeve; *7* – sealing unit body

The volume of interconnecting pores of the circular polymeric T-shaped filler contains up to 30 % by mass liquid metal (Hg, Li, Na). When actuating the system of electromagnets, magnetic flux appears in the seal inducing ring currents in the liquid metallic filler. Under the effect of compressive electromagnetic force, the filler occupies in the seal free volume the position symmetrical to the magnetic poles. When the pressure drop Δp appears in the seal, the liquid metal moves from this position towards one of the magnets. Its magnetic linkage grows, as well as the inductive ring current and respectively the electromagnetic force holding the filler in the pores augment. Therefore, in response to the pressure drop value, the electromagnetic force changes automatically directed opposite to the vector Δp. The experiments have displayed that this seal pressurizes the media, which are both under excessive pressure (2÷3 MPa) and vacuum (to 10^{-4} mm Hg). The electromagnetic interaction of the liquid metallic filler and the field of electromagnets fulfill the FBS function.

The liquid-metallic seals are used for pressurization of the shafts outlet of rotary and reciprocal motion in the vacuum on satellites and space vehicles.

Electrorheological liquids are slurries of dielectric particles of colloid size ($10^{-3} \div 10^{-1}$ μm). The rheological characteristic of slurries (viscosity, plasticity) can be controlled reversibly with the help of strong electric fields ($E > 10^6$ V/m) (Shulman et al 1972). Changes under the field effect of shear viscosity in these systems have been called electroviscous and electroplastic effects.

Electrorheological sealing material (Outland Research, California) represents the elastic matrix with interconnecting pores which are filled with the electrorheological liquid. It changes viscosity in the external electric field within several milliseconds turning the material from soft into elastic. The FBS adjusts the field strength changing the electrostatic interaction of particles according to the effectiveness of seal functioning.

Magnetic liquids are homogenous suspensions of colloid ferromagnetic particles in liquid (water, paraffin, fluorocarbons, etc.). The magnetic permeability of these colloids reaches $\mu \approx 10$, while that of the liquid dispersion is $\mu \ll 1$. The concentrated magnetic liquids have the saturation magnetization $M_S \approx 100$ kA/m at the viscosity comparable with that of water (Fertman 1988). The magnetic attraction force affecting a single magnetic liquid volume in constant magnetic field is equal by value to the product M_S on the magnetic field gradient and the force is directed along this gradient. The non-spherical colloid particles of magnetic liquid are oriented along the force lines of the external field and arrange into dense chains directed along the field. Because of that, the viscosity and strength limit change during the shift of magnetic liquid layer.

The magnetic-liquid seals contain, in the capacity of sealing component, the magnetic liquid interlayer arranged in the friction couple clearance. The seal diagram is shown in Fig 2.8.

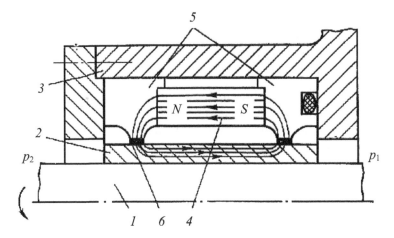

Fig 2.8. Magnetic-liquid seal: *1* – shaft, *2* – ferromagnetic sleeve, *3* – body, *4* – constant magnet, *5* – pole tips, *6* – magnetic liquid

The clearance between ferromagnetic sleeve 2 and diamagnetic body 3 is sealed with circular layers 6 of magnetic liquid. The magnetic flux generated in the magnetic circuit 'circular magnet 4 – pole tips 5 – sleeve 2' is short circuited through liquid interlayers 6 stabilizing their position in the clearance. Under the effect of pressure difference $\Delta p = p_1 - p_2$ the interlayers displace from the equilibrium position, the interlinkage augments and the magnetic force appears directed against the vector Δp. When the pressure difference exceeds the force generated by the magnetic field, breakdown occurs of one of the sealing stages. It is not emergence if the second stage is intact. The next pressure difference reduction compels the smart system 'to heal' because the magnetic field acting as the FBS again collects the magnetic liquid which drops in the clearance, restoring the interlayer (Orlov 1977).

A_7B_{29} are the SM chemically reacting with sealed media under control of the FBS which adjusts the degree of material solidification.

Liquid sealants are viscously flowing mass based on polymers and bloomers, which solidifies in the sealing joint clearance (Pinchuk and Neverov 1995). The thiocole sealants contain the 'latent' curing agent reacting with thiocole (polysulfide rubber) only at the moisture presence. The sealant cures in contact with aqueous sealing media. The FBS adjusts the curing degree in response to the quantity of water medium leaks. The anaerobic sealants are multicomponent compositions based on the polymerizing compounds of acrylic series. When kept in the air, they preserve the original viscously flowing state for a long time but solidify fast at room temperature and oxygen deficit in the clearances about 0.1 mm wide. The non-solidified anaerobic sealants possess a high penetrability, turn after polymerizing into three-dimensional cross-linked polymers and assure joint tightness of mating parts functioning as glue. The FBS adjusts the cross-linking degree in response to oxygen concentration, which reduces when liquid sealing media are delivered into the clearance.

Solid sealing materials are a vast range of substance from plastic compositions, approaching on its properties to the liquid, to the hardest and strongest constructional materials.

The drawback of solid seals as compared with gaseous and liquid ones is the need to fit accurately their contact surface to the surfaces of mating parts for the purpose to assure the least clearance in the joint. It is achieved by finishing the seal mating parts that requires much man hours and results in high cost of these precision parts. The loading of the joint sealing by compression deforms the microroughnesses in the contact and brings the mechanical interface closer. However, it produces intensified seal wear in moveable joints. Nevertheless, the mechanical strength of solid seal materials enables one to use them within a broad range of pressures of sealing media and speeds of rotation of seals moveable parts.

The upper limit of working temperature is ~300 °C for the majority of seals [62]. That restricts the use in contact seals of polymer materials, which continuous service temperature does not exceed 175 °C. That is why composites are often used to fabricate seals, which are based on organic-silicon and heterogeneous-chain polymers reinforced with glass, quartz or carbon fibers. Polyorganosiloxanes, polyphenylons, fluoroelastomers are broadly used as binders having the performance limit, which corresponds to the range 175÷325 °C (Pinchuk and Neverov 1995).

The typical instances of SM for contact seals are given below.

A_1B_4 are the SM in which rheological characteristics are changed purposefully under compression adjusted by the FBS being the property of material structure gradient.

Mixtures of polymers are composite materials comprising of different polymers. Their mechanical properties depend, in the first place, on the thermodynamic compatibility of components, which is an order of 0.1÷1.0%. Therefore, the spontaneous mutual dissolution of most polymers is impossible, and their mixture is a dispersion of one polymer in another. The size of the particles in the dispersed phase in the mixture of polymers depends on the blending conditions and ranges from 0.1 to 10 μm. If the size of particles exceeds the last value, the blending is carried out ineffectively.

Contradictory requirements are put forth in the contacting seals, the main being the combination of high deformability (to fill up clearances in the contact of the sealing conjugation) and sufficient strength (to balance the pressure drop of sealing media). It is possible to combine these qualities provided the deformation-strength characteristics of seals vary according to the distance to the contacting plane. The model of a perfect contact seal presumes the availability of strength gradient, which augments in the direction opposite to the contacting plane with the mating member (Pinchuk 1992). The mixtures of polymers with gradient of the dispersed phase distribution enable to implement this model. The FBS in the smart gradient seal is incorporated into its structure, and its characteristics are determined by the size and distribution of particles in the dispersed phase of polymer 1 in the matrix from polymer 2.

A_3B_{1-4} are the smart sealing materials which coefficient of temperature expansion is optimized by the FBS initiating the structure modification, stresses appearance, rheological processes, sealing element deformation.

Sealing materials for low temperature seals behave at negative temperatures (minus 60÷300)°C. In these conditions, the operation of rubber seals with glass transition temperature minus 63÷25 °C is complicated considerably because of rubber shrinkage. If the rubber seal is installed in the joint of metallic parts at normal temperature T_1, while they operate at the reduced temperature T_2, the rubber high elastic deformation is 'frozen', and the shrinking stresses appear in the seal (Avrushchenko 1978):

$$\sigma = \int_{T_2}^{T_1} \frac{E(T)}{1 - \mu(T)} (\alpha_p - \alpha_M) \, dT,$$

where $E(T)$ and $\mu(T)$ are modulus of elasticity and Poisson coefficient of rubber, α_r and α_m – coefficients of thermal expansion of rubber and metal. The shrinking results in reduction the degree of sealing tightness.

A similar problem is observed in flanged joints of pipelines transporting cryogenic fluids. A temperature drop between flanges and securing bolts may reach 300°C when the cryogenic apparatus starts or stops. The joint gets loose because the bolts are not tightened properly determining joint contact pressure. For instance, if the temperature drop $\Delta T = 150$°C appears in the pipeline 100 mm in diameter, the clearance in the joint to be sealed increases to 60÷70 µm during a few seconds (Babkin et al 1977).

Smart sealing materials should eliminate the gaps with the FBS, which adjusts the parameters of thermal expansion or the shape of the sealing element (the effects of 'shape memory', or auxecity) or initiates the viscous fluid release from the seal into the gaps. The system of heat transfer adjustment with the help of computer ensuring momentary temperature legalization of all parts of the joint at any temperature drop ($\mathbf{A_1B_9}$) seems almost fantastical but realizable at state-of-the-art thermal engineering.

$\mathbf{A_1B_{17}}$ are smart sealing elements in which under load appears the electric current used to adjust the contact pressure in the seal.

Stationary seal of interface between parts from different metals separated by polymeric film or coating represents a galvanic pair generating current (Belyi and Pinchuk 1980). The current value depends on the contact pressure in the seal (Goldade et al 1981). Figs 2.9, *a* and *b* show the design of these seals. The current value registered between the interfaced parts depends on pressure in the contact and seal temperature (*c*). To make this seal smart, it should be provided with computer transforming the electric signal into a command to the device to load the interfaced parts.

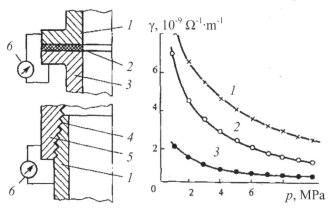

Fig 2.9. Flange (*a*) and threaded (*b*) sealing joints: *1* – pipes from metal-1, *2* – gasket, *4* – coating, *3* and *5* – pipe and clutch from metal-2, *6* – galvanometer; *c* – conductivity dependence of polyvinyl butiral coatings on compression tension at different temperatures (°C): *1* – 80; *2* – 60, *3* –30.

A_1B_{23} are the SM for contact seals which deformation initiates the FBS intensifying diffusion processes in the seal.

Polymeric gel sealing materials are the microporous polymeric matrix in which free volume contains liquid plasticizer (Pinchuk and Neverov 1995). The seal vibration causes the redistribution of liquid in the sealing element filling up the clearances in the contact and heals the defects in the structure by filling them with hydrophobic plasticizer. Thus the FBS is realized optimizing the seal tightness in relation to the water sealing media.

A_6B_3 are sealing elements becoming deformation under magnetic field which is the source of contact pressure.

Magnetoplastics are polymeric composites filled with hard-magnetic ferrite particles and magnetized. They serve to fabricate parts of refrigerators of which doors are magnetically attracted to the metallic body sealing the refrigerating chamber (Voronezhtsev et al 1990). The SM of this class should be provided with the FBS adjusting the force of door attraction without any extra operational and ergonomic factors, such as temperature and humidity of the room and cooled product, door attraction conditions 'open-closed', etc.

A_8B_2 are the SM which swelling in the clearance when contacting with the sealed media, are controlled by the FBS adjusting the swelling degree according to the needed tightness.

Materials based on gel (Filippova 2005) swell differently in the media with different acidity pH. The pressure developed in such sealing element at swelling reaches hundreds MPa. Fig 2.10 illustrates the properties of the materials of this class. One can see that the degree of tightness of the seal with gel sealing element can be adjusted automatically when the composition of the sealed medium changes, even when the loading parameters stay unchanged. Gel materials are used in the oil-producing industry.

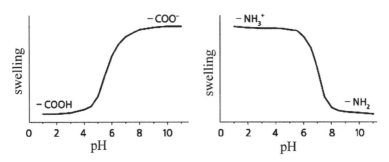

Fig 2.10. Swelling of gels containing weak acid groups in macromolecules (*a*) and weak base group (*b*) in water media of different pH

Swelling packers are the devices lowered on pipes into the bore well to separate the bank from the annular space or two banks between themselves. The packers (elastomer or composite) swell in the reservoir water and in the oil sealing the clearance between the pipe and bore wall. The FBS in the smart system adjusting the well product composition is envisaged in the packer material structure when it is fabricated.

The self-sealing aircraft fuel tanks were developed during WW2. The tank consists of two rubber layers: the external one – from the cured and the internal one (coated inside with fuel resistant varnish) – from the uncured rubber. When a bullet punches the tank, the fuel spills, contacts with uncured rubber which swells and 'heals' the punch. The FBS in this smart system is implemented due the thermodynamic compatibility of the fuel and non-cured rubber.

Many examples of smart sealing materials can be shown because an enormous number of performance factors affecting the contact seals causes still more answers of different nature which put in action the FBS. For instance, the following ways of developing smart seals have been omitted: the freezing seals (Pinchuk 1992); the seals in which thermal expansion optimizes the release of isolating liquid contained by the seal (Pinchuk and Neverov 1995); the stuffing of sealing materials with microcapsules contained such liquid (Goldade et al 2005); electric polarization of sealing materials in operation resulting in creation of mechano-, chemo-, tribo- and magnetic electrets, the field of which reduces the penetrating capability of sealing liquids (Voronezhtsev et al 1990); mechanochemical reactions in moveable seals modifying the surface characteristics of the tight contact (Heinicke 1984), etc.

The sealing problem is most challenging in operation of moveable seals where contradictory requirement of combining high sealing ability and wear resistance is presented. The high sealing ability means when the contact area of interfaced surfaces approaches 100 % causing the growth of friction and friction related heat generation. Vice versa, the clearance formation between friction surfaces brings to wear reduction and friction heating of the sealed interface. The heating has a complex effect on the tightness and wear resistance of moveable seal. This problem can be solved with the cybernetic system capable to comprehend, memorize and use the information about the processes listed above in order to optimize them.

2.3.2   Filtering materials

The filtration is the separation of slurries and aerosols with the help of porous membranes letting though the phase-carrier (liquid or gas), but trapping the distributed solid or liquid particles. When the phase-carrier moves through the membrane, it meets with aero- and hydrodynamic resistance to overcome which the pressure drop should be set on the membrane. The latter is created in the filters or the devices containing a porous membrane or a filtering element (FE) and a unit maintaining the pressure difference from both sides of the FE (vacuum, excessive pressure or combined). FE are fabricated from filtering materials (FM), that is solid materials provided with the system of communicating through channels. The contaminating particles are trapped in the FM by following mechanisms.

The large particles (>20 μm) are trapped by the mechanism of inertia trapping, deposition by gravitation, and sieving effect:

- the inertia trapping of particles carried by the stream of gas or liquid which passes through the structural FM element occurs if the particles under the effect of inertial forces deviate from the stream motion, collide with the structural element and stop;
- the deposition by gravitation occurs under the effect of earth gravity field;
- the sieving effect occurs when the sizes of particles are too large to pass through the FM pores and channels.

The fine contaminating particles (<20 μm) settle due to the mechanism of contact (touch) and diffusion:

- the touch occurs when the particle approaches the obstacle to a distance shorter than its radius and adheres them under the effect of Van-der-Waals forces;
- the settlement by diffusion is typical for the particles less than 1 μm in size accomplishing the Brownian motion, which are trapped by the FM due to the random deviation from the stream direction.

The instances of the SM of this class are given below.

A_3B_3 are smart FM in which the FBS actuates when the temperature changes adjusting the cross section area of filtering channels in the material structure.

The FM with adjustable free volume can be fabricated from chemical fibers which shrink considerably when heated just by 20°C, while they elongate reversibly when cooling (Barukhin and Babkin 1995). The FBS adjusts the FE temperature in response to the filtering medium viscosity changing purposefully the filtration fineness and filtering efficiency, thus simplifying the FE cleaning from trapped solid particles.

A_6B_{18} are smart filters in which the filtering medium stream is exposed to the electric field effect, while the filtration process is optimized by the FBS adjusting the electric polarization and adsorption of trapped particles.

The antimicrobial filters are intended to purify water from microorganisms. Their principle of action is based on the fact that all microbes carry the predominantly negative electric charge (Bunin and Voloshin 1996), therefore, the FM with a positive electrostatic potential, trap them virtually irrespective of the size of pores. The mechanism of the FBS action consists in the expedient adjustment of surface and volume electrocoagulation of microorganisms. The coagulation begins with the electrophoretic movement of biological particles to the positively charged FM parts. Then, as a result of dipole-dipole interaction mechanism, aggregates of bioparticles are generated, which the filter traps easily (Andreev et al 1989).

The Institute of physics and materials science of the Siberian branch of the Russian Academy of Science has developed smart FM 'AquaVallis' based on the aluminum hydroxide having the positive electret charge. It is able to purify water fully from viruses, bacteria and parasites.

The electret fibrous FM based on thermoplastics are intended mainly to purify gaseous media. The fibers are sources of electric field under the effect of which the contaminating particles are polarized and trapped by fibers on the mechanism of electrostatic attraction. These FM combine better filtration fineness with low aerodynamic resistance because the cross sizes of FE through channels can considerably exceed the size of trapped particles. The FE produced by the melt-blowing process have the structure of chaotically distributed fibers, which are bonded by the adhesive links in contact spots. That is why the elements can be shaped and sized as needed during fabrication without any extra processing. The melt-blown FE carry the processing electret charge (Pinchuk et al 2002), therefore, the operation of extra electric charging is unnecessary. The prototype of these SM is the Petryanov FM.

Petryanov FM are fine (~10 μm) polymeric fibers charged while fabricating from the solution. They are applied to the woven or non-woven load bearing substrate (Petryanov et al 1963). During the WW2 they were used intensively to produce gas masks. At present, the Petryanov's FM and their thermoplastic counterparts serve as a base of respirators, i.e. devices for individual protection of respiratory organs from dust and harmful substances. Their advantage is the insignificant pressure drop at the FE, 4÷20 times less than at similar non-electret FE (Kravtsov et al 2003).

The FBS role in the electrostatic filters is fulfilled by different ability of trapped particles to electrocoagulation depending on their size and charge.

This class of smart systems covers the filters in which the so-called exchange adsorption takes place.

The household water purifying filters contain adsorbents which implement the phenomenon of exchange adsorption. The adsorption FE installed in the filter trap the harmful cations or anions located in the water, release an equivalent number of neutral ions of the same sign:

adsorbent⁻(H⁺) + Na⁺,Cl⁻ → adsorbent⁻(Na⁺) + H⁺,Cl⁻

adsorbent⁺(OH⁻) + Na⁺,Cl⁻ → adsorbent⁺(Cl⁻) + Na⁺,OH⁻.

The designations of ions forming a double electric layer on the FE are enclosed here in brackets.

This principle is implemented in the filters 'Smart Water' popular in Europe and USA. They contain the FE from ion-exchange resins of the food class. The FBS function is fulfilled by the electronic block equipped with a diagnostic system, water consumption archive, sound and light alarms.

A_7A_{22} are the smart filters in which the filtering medium is exposed to the magnetic field controlled by the FBS adjusting adsorption, magnetic and electric coagulation of trapped particles .

The device for water magnetic treatment appeared after the invention in 1945 by Belgian engineer T. Vermeiren the method of steam-boilers preservation from sludge deposition. The water pumped into the boiler is passed through magnetic field. As a result, the dissolved salts do not deposit on the heated boiler surface but separate as loose flakes are easily removed by the boiler blowing through (Sokolskii 1990). After that the companies 'Erigo' (Belgium), 'Polar' (UK), 'Packard' (USA) began to produce the devices for water cleaning using constant magnets. In the 1990s the polymeric fibrous magnetic FM were developed consisting of fibers filled with ferromagnetic powders (Pinchuk et al 2002). The FBS function in these smart systems is fulfilled by: i) the capability to rearrange in the magnetic field for water ions gathered round by the 'coat' of water molecules (hydrated ions); ii) the ability of ferromagnetic microparticles to undergo magnetic coagulation; iii) resonance magneto-hydrodynamic phenomena, etc. (Klassen 1982).

A direct proof of the magnetic treatment effect on the structure transformation of hydrated ions are the results of the following experiment (Sokolskii 1990). The half transparent membrane of collodion was mounted tightly in the vessel; separating the vessel in two volumes. The volumes were filled with $CaCl_2$ solution (10 mg-eq./l), one of them – with common solution, another – with magnetized solution. The current between electrodes dipped into the vessels was registered. The currents were found to be different when the electrodes were transposed. It is due to the fact that the original hydrated ions Ca^{2+} pass harder through the membrane pores than the same ions of magnetized water which 'coat' became less dense. The magnetic field effect intensifies considerably if water contains ferromagnetic iron oxides.

The vehicle oil filter protects the oil lines from impurities and oil decomposition products. The clean oil determines the quality of lubrication of moving interfaces, as well as cleaning and cooling of cylinders from wear products and incomplete fuel combustion. The oil traps these impurities and particles of dust getting into the engine through the air filter, and carries them to the crankcase sump where the FE is installed. The problem solved at the oil filter development is to find a rational relation between the filtration parameters: the finer the filtration, the less is the filter efficiency. The above-mentioned melt-blown FM can be provided with the property of being the magnetic field source. To this end, the polymeric fibers are filled up with magnetic solid particles and magnetized in the viscous flow state to facilitate the orientation of particles by the axes of easy magnetizing in the direction of the external magnetizing field. These FE with low hydrodynamic resistance trap even the non-magnetic contaminating particles because of the FBS implementing the mechanisms of magnetic coagulation, adsorption and magnetic attraction of coagulants. The smart magnetic oil filters do not lose their performance even after the vehicle runs almost 100 thousand kilometers (Pinchuk et al 2002).

A_9B_{32} are the smart filters in which the contaminants of filtering fluid are eliminated by the cultivated in the FE microorganisms adjusted by the FBS which function is implemented by the inherent ability in microorganisms to control the kinetics of biochemical reactions of metabolism expediently.

Biological filters (*biofilters*) are the structures for biological cleaning of wastewater. The tank of a biofilter is filled with granular FM (loading), which is populated by bacteria destroying organic impurities in wastewater. The impure water is pumped into the tank where it is filtered through the loading leaving on the surface a dirty film utilized by bacteria. In such a way, natural biocleaning is implemented in industrial production as in nature.

The first biofilters appeared in England in 1893. By middle of 20th century the biofilters lost their appeal drastically in favor of aerotanks in which the wastewater is cleaned by oxidation of organic impurities with microorganisms living in the layer of silt at the bottom of the aerotank. In 1960-70s, when the ecological problem showed aggravation and new highly effective polymeric carriers of microorganisms were developed, the biofilters regained their broad dissemination.

The technique of melt-blowing permits to produce fibrous electret and magnetic carriers of biomass with optimal shape for all systems of biofilters. The biofilters with such loading reach the stationary mode quicker as compared with counterparts (claydite) – from 30 to 14 days. They demonstrate higher and more stable parameters of biocleaning, particularly at the stage of startup and under peak loads, in other words, at the elevated contaminant concentration and toxic effects on the biofilm. Weak magnetic fields of fibrous biomass carriers ($B_r = 0.1 \div 0.5$ mT) promote the accelerated proliferation of microorganism carriers and prolong their metabolic activity [80].

As a smart ecological system, the biofilter is notable for stable equilibrium, in other words, it is able to restore by self-adjustment the optimal performance and effective cleaning after deviation from stable state to overloading in operation. The self-adjustment is initiated by the microorganisms modifying the kinetics of biochemical reactions of metabolism expediently.

Summarizing it is noteworthy that filtering systems are notable for multifactorial operational effects and variety of responses (reactions) of various natures from the side of FM. It creates prerequisites for direction a part of filtering energy to the FBS functioning and optimizing this process.

2.3.3. *Heat-insulating and sound-absorbing materials*

The heat insulation is protection of buildings, heat generators, refrigerating chambers, pipelines and other objects from negative heat exchange with environment. The heat insulation is provided with heat insulating materials. The sound-absorbing materials absorb noise in the room and/or protect it from sound penetration from outdoors. In most cases the heat-insulating materials can also absorb noise, that is why they are often interchangeable and one material can serve the double function of heat insulation and sound absorption.

The heat-insulating materials are characterized by high porosity and low heat conductivity. *Inorganic materials* of this class are manufactured from mineral stock (cement, glass, glass fibers), rocks and minerals (perlite, vermiculite, diatomite, asbestos, limestone, gypsum, etc.). The instances of inorganic heat-insulating materials are foam glass, porous concrete, blister perlite. *The organic materials* are products of processing wood, peat, and gas filled plastics. The heat-insulating *materials of mixed types* are the mixtures of mineral binding matters and organic fillers.

The materials of this class are divided into heat-insulating proper and heat-protective. The former are intended to lower the objects heat losses. That's why they should have a minimal heat conductivity coefficient. The latter are intended for personnel and equipment protection from undesirable external heat effects. They should have as little as possible product of the heat conductivity coefficient by the material density (Guzman 1995).

The heat-insulating SM adjust the temperature of the medium in which the article is kept, purposefully correcting the natural changes in the own heat physical characteristics with the help of the FBS consuming external heat energy. To show instances of these SM is rather difficult, most of heat-insulating materials should be referred to the multifunctional and adaptive. To provide them with the FBS equals the development of the materials of future.

A_3B_2 are the SM heating of which switches on the FBS are healing the structural heat damages.

The heat-protective coat of spaceships should not only be filled with sensors tracking temperature, pressure, flight velocity, etc., but also should incorporate the components which heat destruction would release the agency of damaged coat parts healing to be accomplished by the command of sensors.

A_3B_3 are the SM experiencing heat deformation which the FBS directs to stabilize the operating temperature conditions.

Heat-insulating foams and resins are filled with microcapsules containing fire retardants. In case of fire, the microcapsules break, the fire retardants evaporate and form the atmosphere inhibiting flame propagation (Guzman 1995). These adaptive materials are advisable to provide with the FBS, which would redistribute fire retardants in the volume of material replenishing them in the coat surface layer.

A_3B_{10} are the SM in which, when the temperature varies, the FBS initiates the set-up structural changes.

The smart dress, in particular, *environmental suits*, contains heat-protective elements from fibrous materials. When the temperature varies, their volume changes optimally due to the heat deformation of fibers, thus modifying both the heat conductivity coefficient and the density of elements adapting to the heat-protective suit characteristics. The DuPont Co implements a special program of developing smart textiles.

The refrigerator units are smart heat-protective systems providing continuously artificial cooling of the refrigerator chamber by heat elimination. In the simplest case, circular reversible thermodynamic cycles evaporation/condensation of refrigerants are accomplished. Refrigerants are easily evaporating low boiling fluids – NH_3, propane-propylene mixtures, freons, liquefied air, N_2, He, etc. They receive the heat from the refrigerator chamber at a lower temperature and return it to the heat receiver at a higher temperature and restore the original state (Ivanov 1999). The FBS function in these systems is performed by natural adiabatic processes of refrigerant evaporation and condensation.

A_4B_3 are smart heat-protective systems in which, when heated, the components of protective material melt actuating the FBS which switches on cooling.

Composite materials with easily melted components, which fluxion signals about the critical state of heat protection, can serve as elements of the smart system. The fluxion of components-sensors serves as a signal to actuate the FBS which switches on the cooling unit of the heat-insulating screen.

A_3B_{11-12} are the SM the operation of which at temperature variations maintain the feedback initiating the melting or solidification of low melting components.

The plaster-thermostat maintaining constant indoors temperature contains 10÷20 % plastic microcapsules filled with paraffin. When the temperature in the room exceeds the paraffin melting temperature (24 °C), it melts and absorbs the heat of the room-heated air. When the room cools, paraffin solidifies releasing the latent heat of melting (Barukhin and Babkin 1995).

A_4B_{19} are SM adjusting the intensity of passing through heat flow with the help of FBS changing the type of material conductivity.

The thermostatic window glass has a vanadium dioxide lamina, which displays the properties of metal or semiconductor at temperature variations within 30 °C. In the metallic state, the coating reflects the infrared component of solar radiation well preserving the room temperature from heating. When it becomes cold, it switches over to the semiconductor state and admits the heat emission.

These instances evidence that the range of smart heat-protective materials is still very narrow, apparently due to the fact that the main requirement for them is a high reliability. The simple materials satisfy this requirement to the most degree. The most evident tendencies of creating SM of this class are the following: i) adjustment with the FBS the structural state of materials and initiation of physical and chemical processes of self-adjustment of the structure at temperature drops; ii) development of composite materials with components being heat sensors which signal actuates the FBS.

The sound-absorbing materials transform the sound waves energy into heat energy. They can be divided into two groups: i) noise-absorbing materials which are intended to reduce noise in the premises and ii) sound-insulating ones intended to insulate structures from sound waves generated in the air during impacts and constructions vibrations.

The main characteristic of the materials of this class is the sound-absorption coefficient, which is the relation between the energy absorbed by the material specimen and the total sound energy received by its surface:

$$\alpha = (Z_a - 1)/(Z_a + 1),$$

where $Z_a = Z_0$ cth $\gamma_0 d$ is the specimen acoustic resistance, $Z_0 = p/v$, p – sound pressure in the running wave, v – the velocity of sound oscillations at any point of the endless medium, γ_0 – the constant relating to the material density and porosity, d – the specimen thickness (Borisov 1990).

Noise-absorbing materials used in the building industry most often represent the densified linen from glass and mineral fibers (basalt, granite, diabase, limestone, shale). The density of these materials is $\rho \leq 70$ kg/m^3, the sound-absorption coefficient is $\alpha = 0.70 \div 0.85$ within the frequency band 500÷1000 Hz. Sometimes the fiber layer is placed between perforated sheets.

Sound-insulating materials are predominantly used as fillers of voids in multilayered floors and partitions. They are characterized by low values of dynamic elasticity modulus E_d. The fibrous materials in the shape of rolls and mats have $E_d < 1.0$ MPa and $\rho = 75 \div 175$ kg/m^3. The materials from porous-spongy foam plastic or porous rubber with $E_d = 1 \div 5$ MPa are used as gaskets. The dry substances (cement or metallic slag, bloating clay aggregate, sand) with $E_d = 5 \div 15$ MPa are used as bedding in layers between floors (Borisov 1990).

The mechanism of energy absorption of sound waves by these materials is the following. In homogeneous porous materials, the wave energy is consumed: i) to overcome viscous friction the air stream meets in the pores of material, ii) for heat exchange between wall pores and air, iii) for processes of relaxation in materials of non-perfect elasticity. The sound absorption augments with increasing: materials porosity (no less than 75 %), pores extended surface, number of communicating pores. The through communicating and ramified pores are preferable with the diameter $0.1 \div 1.0$ mm (Isakovich and Osipov 1970).

We could not find the instances of SM adaptable to acoustic effects. Whereas the physical mechanisms of sound absorption and the structure of used sound-insulating engineering materials, it can be assumed that it is advisable to maintain the optimal value of coefficient α by adjusting the acoustic resistance Z_a of sound screen. The most apparent way of solving this problem is to change purposefully the material density and porosity on which the constant γ_0 and screen thickness d depend.

A_4B_3 is a sound-absorbing system consisting of acoustic screen provided with sound pressure sensors. They generate the signal switching on the FBS which actuates the unit of screen mechanical deformation. It determines the adjustment of channels flow section, the screen thickness and density.

The sound absorption coefficient of foam polyurethane specimens is known (Adintsova et al 1996) to undergo the maximum in response to the applied compressive stresses (Fig 2.11). Adjusting the latter, it is possible to achieve that the screen would show the maximum α at any variations of sound oscillations.

This taxon covers the smart system like *the sound absorbing screen* fixed on tight elastic ropes. The sound waves exert pressure on the screen transferring them to the spring-tensioned rope structure. Its vibration amplifies the sound absorption. If the frequencies of sound wave and rope vibration coincide, the resonance appears at which the losses of sound energy augment drastically suppressing the sound. The automatic unit adjusting the ropes tensioning and vibration frequency depending on the sound wave frequency implements the FBS function in this smart system.

A_4B_{10} are smart sound-insulating systems in which the sensor of sound pressure sends a signal and actuates the FBS adjusting the temperature of

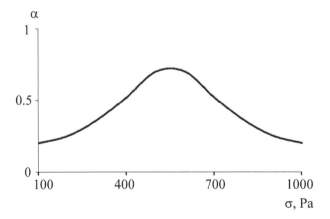

Fig 2.11. Dependence of absorption coefficient α of sound waves (v = 500 Hz) in foam polyurethane specimen on applied compressive stresses

the sound absorbing screen. At temperature 'smart' variations, the parameters of heat exchange between pore walls and air change optimally, as well as the rheological parameters of walls. It results in sound wave energy consumption for viscous friction of the air stream in the pores, as well to increase sound resistance of the protective screen.

2.3.4. Radio wave absorbing materials

The absorption of radio waves is the conversion of energy of the electromagnetic wave propagating in the medium with the frequency $10^5 \div 10^{12}$ Hz into other energy types. The non-resonance radio wave absorption is the wave energy conversion into heat energy in the medium. The radio wave energy, at resonance absorption, is consumed for transition molecules of the absorbing matter into higher energy states.

The radio waves absorbing materials (RAM) absorb streams of electromagnetic energy with density $0.1 \div 8.0$ W/cm^2 of the radio-frequency bandwidth at their minimal reflection ($0.001 \div 5.0$ %). Propagating in the RAM, the electromagnetic emission (EME) generates the alternating field which energy in the perfect case dissipates (scatters) virtually fully. In addition to dielectric and magnetic losses in RAM, other phenomena occur in this materials, such as dispersion, diffraction, interference and full wave internal reflection, which cause extra attenuation of EME energy due to the Raleigh dissipation, summation of waves in the counter phase, etc. (Batygin and Bravinskii 1995).

The RAM as coatings on the metallic bodies of aircrafts, tanks, missiles and ships are used in order to mask them against radars and to protect the people against radiation of high intensity. The coatings improve the performance of antenna arrangement, ensure the energy absorption in electronic circuits and radiation compatibility of equipment parts. As the electromagnetic technologies are progressing, the problem of radiation absorption turned out the global problem of human electromagnetic safety (Grigor'ev 2002).

There are no universal RAM equally effectively absorbing the EME of any frequency. Therefore, it is impossible to make the object fully invisible to radars, it is possible only to reduce considerably the dissipation effective surface for the waves of certain frequency. It does not make the object invisible at other frequencies proving that the smart RAM are still not created. The wonderful achievements of radio absorption technology like 'Stealth' permitting in many cases to have invisible reconnaissance aircrafts, submarines and other objects of military equipment, do not disprove this statement.

Nevertheless, the state-of-the-art of radiophysics, materials science and chemical technology is sufficient to equip the RAM with feedback. Let us see the following instances.

The resonance RAM are applied as coatings on reflecting surfaces of the object to be masked. The coating thickness should correspond to one fourth of the wavelength of radar station emission. The high frequency emission is reflected by the external and internal surfaces of coating implementing the interference mechanism of original wave neutralizing. As a result, the energy of incident emission is absorbed. The efficiency of resonance RAM is restricted by a narrow band of emission frequencies.

The problem is to develop a FBS changing the distance between EME reflection planes respectively to the incident emission wavelength. The virtual SM with such FBS corresponds to the taxon A_4B_3.

Non-resonance RAM are plates filled with soft magnetic ferrites and metals. The remagnetization of ferrite particles and appearance of eddy currents in metallic particles under the effect of high frequency EME leads to conversion of emission energy into heat. The advantage of these RAM is broadbandness – the EME absorption within a wide band of frequencies. The heat released during absorption is advisable to use in the FBS generating a field in the electromagnetic screen (EMS), which interacts with the incident EME, for instance, by implementing the Peltier effect (A_4B_{17}).

The gradient RAM are filled or sandwich materials which dielectric or magnetic permeability is not constant, but varies in dependence to the distance to the plane, on which the EME falls. The screens from these materials are characterized by the minimal radio waves reflection and their absorption grows as they penetrate deeper into the material. The evident way of rendering the latter with the properties of artificial intelligence is the screen equipping with the FBS changing the absorption gradient of the incident EME (A_4B_{17-18}).

The interference EMS is of the form of coating comprising of alternate dielectric and conducting layers deposited on the rigid substrate. This structure assures the phase shift of the wave reflected by layers to almost 180° and its attenuation. Certain compliance is needed for this between the incident wavelength and radio physical characteristics of coating layers. The interference screens need the FBS, which would adjust the electromagnetic characteristics of layers in response to the wavelength of the incident EME. The phase shift of the reflected wave should be 180° at any wavelength.

The dissipating RAM assures multiple reflection and dissipation of radio waves. The walls of the radio-tight echoless chambers are faced with cellular structures comprising of plastic hollow elements of pyramidal or spike-like shape (Fig 2.12). The inner walls of the elements *a* and *c* are covered with graphite, while the outer walls of the element *b* – with the layer SiC (Batygin and Bravinskii 1995). The panels of this design are believed the best based on the criterion of the maximum EME absorption and minimal reflection. A smart screen design can be imagined equipped with a drive changing the size and shape of the elements by the FBS signal in response to the frequency variations of the incident emission.

| *a* | *b* | *c* |

Fig 2.12. Radio-absorbing panels of echoless chambers with elements like: *a* – moss, *b* – sedge, *c* – lotus

2.3.6. *Anticorrosive coats*

Corrosion is the destruction of condensed bodies induced by chemical and electrochemical processes of interaction with environment. The corrosion originates on the body surface, therefore, the technique of surface protection is extremely challenging.

The participants in the corrosion process are the material and environment. Therefore, to inhibit corrosion, the article surface layer should be modified, the medium aggressiveness should be reduced (to remove all corrosive components – deoxidizing, neutralizing; to introduce into the medium the substances inhibiting corrosion – inhibitors, passivators, etc.) or generally the material should be isolated from the medium. All these methods can be implemented by multifunctional coatings, which creates favourable prerequisites to impart them the attributes of artificial intelligence. The characteristics of corrosion-resistant coatings of the attractive feature is the least materials consumption, as well as examples of smart coatings on metallic articles are made, are given below.

Metallic coatings are among the most used in the engineering of protective means against corrosion. These coatings are multifunctional: they can add to articles extra hardness, wear resistance, electric conductivity, reflecting capability, and improve the appearance of articles. The metallic coatings are divided into anode and cathode types depending on the mechanism of protective effect.

A_8B_{18} are the smart metallic coatings containing the FBS which, when 'attacked' by the environment, electrically polarizes the metallic substrate.

Anode coatings have a larger negative electrochemical potential than that of the protected metal. Like all coatings, they prevent the contact of metal with environment, but even the non-continuous anode coatings protect the metallic substrate electrochemically: in contact with electrolyte they become anode with the substrate – cathode which is not subjected to dissolving until there is electric contact of coating with substrate and there is current between them. The FBS function in this smart system is fulfilled by substrate electric polarization. The anode coatings on steel articles are made from Zn and Cd.

Cathode coatings have the electrochemical potential more positive than the metallic substrate. They create an impermeable barrier to corrosive medium which operates until its continuity is upset. If a defect appears in the coating, then a galvanic element appears in which the substrate becomes anode and destroys electrochemically. The defects in the substrate provoke pitting corrosion. The cathode coatings on steel articles are the coatings from Cu, Ni, Sn, Pb, Cr. They need the FBS, which would 'switch on' the mechanism of defects healing in the coating, for instance, by filling them up with corrosion resistant products forming plugs impermeable for the medium.

Thermodiffusive coatings are formed by diffusive saturation of the surface layer of the metallic article with the atoms of other elements: Al (aluminizing), Cr (thermal chrome-plating), Si (thermal siliconizing). The diffusive atoms form in the surface layer resistant to corrosion new phases of chemical compounds or solid solutions. The idea of creating a FBS 'repairing' damages of these phases at the expense of energy of electrochemical interactions, so far seems utopia.

Non-metallic inorganic coatings are characterized by the high chemical stability. The most popular ones are the corrosion-resistant coatings based on the glass enamel, cement and so-called conversion coatings.

The glass-enamel coatings are applied on the articles from cast iron, steel, aluminum and alloys of light metals by the slip technique and fixed by baking. They are characterized by strong hardness, resistance to corrosion, wear and heat. Though the glass-enamel coatings are applied in two stages (light-melting priming and heat-resistant covering layer), they are very vulnerable to impact and heat shock loading. It is hard to imagine that in the FBS, which would heal the cracks in the coating due to impacts. The FBS of rheological nature is more realistic dumping the peak stresses in glass-enamel coatings. These SM correspond to the taxons $A_{1,3}B_4$.

Cement coatings the steel elements of constructional structures protect from corrosion passivating them due to a high pH (~12) of the cement capillary moisture (Neverov et al 2007). The heavy damages of cement layer are self-healing by the products of corrosion. It is an attribute of the A_8B_{29} class of SM: the FBS actuates in the cement layer correcting the response to the environment chemical effect with the help of electrode reaction.

Conversion coatings (phosphate, chromate, oxide) are applied to steel articles by chemical reactions. Their broad proliferation in mechanical engineering is due to the produceability and good adhesion to substrates. These adaptive coatings need a FBS which, when necessary, would provide the protective capability with the help of secondary chemical reactions (A_8B_{29}).

Rust transformers are adhesive priming materials, which turn corrosion products on article surfaces into a layer of insoluble compounds protecting articles. Most transformers contain the ortophosphoric acid turning the rust into insoluble iron phosphates $Fe(H_2PO_4)_3$ and $Fe(H_2PO_4)_2$. These active layers serve as components of smart corrosion-resistant coatings.

Steels producing the barrier against corrosion (Korrosion Träger Stahl – KTS) have been in use in Germany since 1965 as the material for constructional structures (Poller 1982). Instead of the loose layers of iron oxides, a dense protective layer appears from alloying elements resisting to further corrosion. It is because, in addition to the main steel components, *KTS* contains P (0.7÷0.15%), Cu (0.25÷0.55%), Cr (0.50÷1.25%), Ni (0.65%). The costs of production of these SM (corresponding to the taxon A_8B_{29}) are compensated quickly because there is no need to protect structures against corrosion.

Organic coatings, as a rule, possess strong chemical resistance and are characterized by the possibility to introduce active components, which adjust the kinetics of chemical reactions in coating defects.

Rubberized coatings (rubber and ebonite) are applied to protect steel parts from corrosion, cavitation and erosion damages of chemical equipment, dredger parts, propelling screws, etc. (Penkin 1977). The process of 'Liquid Rubber' coatings application is broadly applied in Russia and elsewhere (Boskolo 2006). These coatings are quite thick (centimeters), and it is better to call them facings. The protective ability of multifunctional rubber facing depends considerably on the strength and wear-resistance. Therefore it is advisable to set up a gradient of mechanical properties in the rubberized coatings adjustable by the FBS corresponding the class $A_{1,8}B_{29}$.

Paint-and-lacquer coatings are intended to protect metals against corrosion, would –against humidity, and to add special properties to surfaces (electric and heat insulating, decorative appearance, etc.). The barrier properties of coatings are easily added with the mechanism of sacrificial protection. Chromate and phosphate pigments are introduced into oil paints and organic enamels. They saturate the aggressive liquid diffusing into the coating with ions, which passivate the steel substrate. The degree of passivation in the smart coating is adjusted depending on the kinetics of diffusion processes (the taxon A_7B_{29}). Similarly, the paint-and-lacquer coatings filled with powders of metals with electrode potential less than that of the substrate (for coatings on steel they are Zn, Al) are provided with the FBS 'switching on' the mechanism of sacrificial protection (A_7B_{18}).

Coatings – corrosion indicators based on the transparent acryl varnish contain the phenolphthalein, which paints into the pink color in the alkaline medium. When applied to the aircraft fuselage, these coatings acquire the coloring when the corrosion of aluminum alloy begins accompanied by the appearance of alkaline ions. The painting intensity of the smart coatings is stronger as corrosion is more intensive (the taxon A_7A_{29}) and enables one to detect the corrosive defects on the fuselage when they are about 15 μm deep [94].

Inhibited polymeric coatings have the gel structure on the base of thermoplastics, which liquid phase is the corrosion inhibitor (CI) solution in the plasticizer of the thermoplastic (Pinchuk and Neverov 1993). The gel is a non-equilibrium system, therefore, stresses appear in the polymeric matrix forcing out the liquid phase – so called 'syneresis'. Due to this, the CI delivery from the coating volume to the coating-substrate boundary continues for years. The CI delivery can be boosted by dissolving them in low-viscous transporting fluids. Fig 2.13 shows that the mixtures of CI with the mineral oil are delivered from the coating based on polyethylene (PE) quicker than each component in the mixture separately staring with the liquid phase concentration 25÷30 % in the coating (Neverov et al 2007). The feedback function in this smart coating is fulfilled syneresis which intensity corresponds to the variations of temperature-moisture conditions of coatings operation.

The inhibited polymeric coatings of new generation are 5÷10 μm thick, they comprise the polyelectrolytes (the polymers which macromolecules contain ionogenic groups) and the CI. They are obtained by layerwise deposition of components on the metallic substrate in the electrostatic field. The substrate from aluminum alloys is pretreated by intensive ultrasound emission in the water bath (Andreeva et al 2008). This treatment improves the substrate wetting with the polyelectrolyte and boosts coating adhesion. The protective ability of coatings is adjusted by the FBS

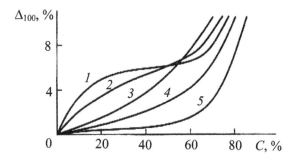

Fig 2.13. The liquid mass Δ_{100} released from PE-based coating per 100 days depending on original liquid phase content: *1* – mineral oil; *2* – inhibitor VITAL; *3* – oil + VITAL (1:1); *4* – oil + MSDA (1:1); *5* – inhibitor MSDA

initiating the substrate passivation by the inhibitor, stabilization of the pH at the coating-substrate boundary by the polyelectrolyte layer, self-healing of defects in the coating due to the mobility of polyelectrolytes chains.

Protective materials are the components of the system 'environment-SM-protected object' in which it is considerably easier to create the FBS than in a separate material. Nevertheless, the demonstrated review proves that there are still few smart protective systems used in engineering. The number of examples of operating smart protective materials we tried to compensate by possible trends of FBS developing in these materials were insufficient. No doubt that these trends will be in progress by all means.

2.4.   Materials for electrical engineering and electronics

Modern electrotechnical materials and materials for electronic equipment are characterized by active response to the electromagnetic field effect. Quite recently, the materials applied in other industrial branches have been used as electrotechnical ones. But with time, the specific power, working voltages and operating temperatures of electrical machinery and equipment grew significantly. The electronic equipment has appeared, i.e. the instruments and devices for processing and transmission of information, computerized systems of management production processes, etc. The materials of this class became to be operated at the cryogenic temperature approaching to absolute zero and also in case of drastic temperature changes ('thermoimpacts'). The phenomenon of superconductivity has become a technical reality. Concurrently the requirements to the reliability of electroengineering devices and radioelectronic equipment have risen tremendously, determined, to a large extent, by the reliability of their electrical insulation, contact connections of semiconductor and dielectric circuit components. The radioelectronic equipment has been developed, in other words, the radio engineering systems and systems of telecommunications containing a huge number of components which production needs special materials. As a result, the electronics became an engineering domain in which the smart materials and systems have for the first time found the real implementation.

To imagine the tendencies of development of smart systems in electronic engineering, let us briefly retrospect the progress of this domain. The arsenal of means of modern electronics has been created during the last several decades. At the beginning of 20 century mankind has done without radio at all, until the middle of the 20th century there has been no television, or computers. The radio engineering has begun to use the electronic instruments after British physicist D. Fleming invented in 1904 the double-electrode valve (the diode) with the red-hot cathode. The diode was developed to detect high-frequency oscillations. After the American engineer L. Forest introduced in 1907 the controlling grid into the valve, the latter became as the three-electrode one. An opportunity has appeared to control the current flowing between cathode and anode. It has enabled to amplify the electric signals. By mid 1930s the valve electronics was mainly developed.

The trend in electronics of developing special instruments handling the electromagnetic waves of super high frequencies (SHF) was no less important. In 1939 the first instruments appear to magnify and generate the SHF oscillations – drift klystrons, vacuum smart triodes with flat disk electrodes and magnetrons to generate powerful SHF oscillations were designed.

The semiconductor electronics was developing intensively in the 1930s. The physical processes in semiconductors were investigated, their thermoelectrical and photoelectrical properties were established, the quantum theory of semiconductors was developed, the theory of 'electron-hole' couples generating was created. The theory of semiconductors developed by the Russian school of Academician A.F. Joffe was corroborated experimentally.

The invention of the transistor in 1948 by American physicists D. Bardin, W. Brattine and W. Shokly initiated the era of radioelectronics miniatuarization. The replacements of lamps with transistors allowed to reduce considerably the weight and size, save the energy consumption and promote the equipment reliability. The perfection of transistor fabrication technology resulted in the appearance of microelectronics. In the 1960s the integral circuits (IC) appeared with all components communicated both technologically and electrically. The achievement of microelectronics can be considered the 'smart dust' or the net of wireless components from a millimeter to dozens micrometers in size which were named *motes* (dust particles). Each mote has an electrical circuit, power supply, sensor and computing unit. The expedient grouping of motes permits one to create flexible nets of management systems (Sailor and Link 2005).

The functional electronics came into life in the 1980s implementing the ascertained equipment function without applying standard components (diodes, resistors, transistors, etc.) based directly only on the physical phenomena in the solid body. The smart systems of functional electronics employs the optic phenomena (optoelectronics), the interaction between the electrons flow with acoustic waves in the solid body (acoustoelectronics) and a number of others.

A new trend of electronics – nanoelectronics – appeared after the atomic force microscopes were created and enabled both to watch atoms and to manipulate them. The nanotechnologies permit to design the IC by placing the needed atoms and atomic structures in a strictly determined place.

The present paragraph deals with the main classes of materials in electroengineering and electronic engineering which comply with the categories of smart materials and systems.

2.4.1. Conductors

Conductors are the substances conducting the electrical current, in other words, possessing high electrical conduction. The conductors cover metals, electrolytes and plasma. Metals and carbon (the conducting modification) are called the first class conductors, electrolytes – the second class conductors. The electrical engineering and electronics use broadly resistive materials – the conductors with a high specific resistance based on metal oxides.

A_3B_{17} are the SM which electrical conduction changes under the temperature effect adjusted by the FBS checking the active conductor resistance to electrical current.

Posistor is the resistor which electrical resistance augments in spurts as the temperature grows. The resistance jump by several orders of magnitude occurs within the temperature range from 70 to 150 °C (Fig 2.14). The FBS originates due to reversible phase transformation of the material from which the posistor is made. Due to this, posistors are used as self-stabilizing heating elements. From the beginning, the posistor operates as a usual heating element. When the temperature of phase transition in the SM is reached, the posistor resistance augments sharply and the current falls to the value insufficient for further heating. When the posistor cools down, its resistance diminishes, the current grows and the temperature reaches again the value corresponding to the phase transformation. The posistor heats quickly, the temperature is maintained very precisely, and as a result, the necessity falls away from complicated and not always reliable computerized management systems. The smart posistor manages itself independently.

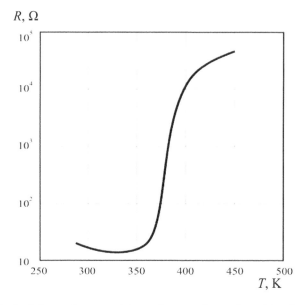

Fig 2.14. Typical dependence resistance-temperature of posistor based on lead ferroniabate (Smyslov 2002)

In Germany, about 50 % of all applications for motor vehicle heating devices fall to the share of posistor heaters (Smyslov 2002). The posistor fulfills the function of temperature controller in the passenger compartment when the engine is not warmed-up enough. The posistor becomes disconnected automatically as soon as the engine cooling liquid temperature reaches the needed level.

'*Perpetual fuse*' is the promotional and jargon name for electrical fuses based on the smart current conductive polymers which like the posistor have positive temperature coefficient of resistance (Kuryshev 2001). During electrical overloading or short circuiting, they behave as usual fuses changing from the low- to the high-resistance state. The resistance of polymeric fuses constitutes only several milliohms in the current conducting state. The self-restoring polymeric fuses are perfect components of the systems, the main requirement of which is the failure-free functioning. They are intended for prolonged application because they destroy at the current many times exceeding the operating current.

The SM of such fuse has the structure of amorphous-crystalline polymer filled with current conducting particles (usually ash). A sharp resistance change of the fuse is due to the phase transformations in the SM. In the 'cold' state, the current conducting particles are localized in the amorphous regions between the crystalline formations, and they connect with each other when their concentration is sufficient. At the polymer melting temperature (the overloading by current), the crystallites melt, therefore the amorphous phase grows in volume. The three-dimensional current conducting structure consisting of filler particles disintegrates leading to the fuse resistance growth. The current reduces to the values which are safe for the circuit elements. The fuse automatically reverts to the initial (working) conducting state when the overloading current is eliminated and the fuse temperature drops. The smart fuses are used in household electronic devices and personal computers to protect them against overloading or short circuit in USB-circuits, FireWire-ports, and other interfaces with power input.

The phenomenon of superconductivity at cryogenic temperatures is used quite extensively in engineering. The superconductivity is inherent to 26 metals. Most of them are first-kind superconductors with the critical transition temperature $T_c \leq 4.2$ к. It is natural that such superconductors are hard to employ in the electrical engineering. Still 13 more elements (silicon, germanium, selenium, tellurium, antimony, etc.) manifest superconducting properties at high pressure. The metals are devoid of superconductivity which are the best conductors in normal conditions – gold, copper, silver. Their low resistance manifests weak interaction of electrons with the lattice. Therefore near the absolute zero point, there exists no attraction between electrons sufficient to overcome the coulomb repulsion, and no transition into the superconducting state takes place.

In addition to pure metals, many intermetallic compounds and alloys possess the superconductivity. The total number of the superconductors known so far is about two thousand. The alloys and compounds of niobium among them have the highest critical temperature. Some of them are allowed to use a cheaper coolant – liquid hydrogen – instead of liquid helium to reach the superconducting state.

A_3B_{19} are smart conductors of which temperature change of electrical conductivity is adjusted by the FBS controlling the conductivity type.

Cryotron is a miniature electronic switch in which the principle of action consists in the jumping of superconductivity state of the main current channel under the action of the controlling magnetic field. Fig 2.15 shows the diagram of the film cryotron.

At the temperature $T<T_c$ gate film 2 remains superconducting until the magnetic field generated by the current flowing through controlling (also superconducting) lead film *1* stays below the critical value for tin. Then, film 2 passes into the non-superconducting state, the current through it reduces sharply. It signals to the FBS to actuate and adjust the current through film *1*. The computer controls the FBS optimizing the magnetic field parameters under assigned program. That is why the cryotron can

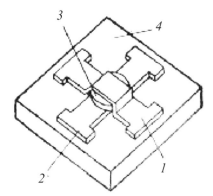

Fig 2.15. Diagram of film cryotron: *1* - controlling film of lead; *2* – gate film of tin; *3* - insulating layer; *4* – substrate (Rakhubovskii 2004)

be attributed to another taxon – A_6B_{17} integrating the smart systems vulnerable to magnetic fields and provided with the FBS which control the reaction of the systems by onset of current.

The cryotrons are used, in the first place, in the logical elements and computer memory cells. The advantages of the cells on the film cryotrons are their speed, meager energy losses and compact size. The instruments incorporating the cryotrons permit one to register fine physical effects and to process a lot of information.

The superconducting electrical cable, first, assures one the least losses in power transmission lines of direct or alternating current, second, limits the unwanted or dangerous shorting currents (Hiroyasu and Masanobu 2006). It is a multilayered design consisting of the central metallic core and two layers of superconducting conductors (Fig 2.16).

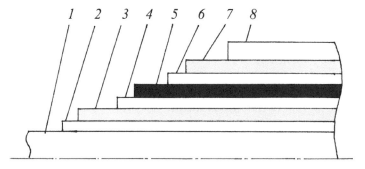

Fig 2.16. Longitudinal section of superconducting cable (Hiroyasu and Masanobu 2006): *1* – metallic core; *2, 4, 6* – insulating layers; *3, 7* – superconducting layers; *5* – normally conducting layer; *8* – protective sheath

Central cable core *1* is formed by twisting multiple normally conducting wires. Two superconducting layers *3* and *7* are separated by normally conducting (copper) layer *5*.

In case of circuit shorting, large currents heat the cable, therefore, the mechanism of superconductivity ceases to be in action. The short-circuit currents are channeled into the central core and copper layer *5* which are electrically connected to the superconducting layers at both cable ends. Normally-conducting layer *5* has the cross section area and the inductance larger than screening superconducting layer *7*, as a result, first, the temperature growth is suppressed in case of short circuiting, secondly, losses of alternating current are reduced when the cable operates normally. Thereby, the electric FBS operates in the cable. Its mechanism consists in the conductivity changes of cable layers: if the temperature rises due to short circuiting, the current begins to flow along the normally conducting central core and the copper layer; as a result, the cable temperature drops, reaches the critical value and the current returns into the superconducting layers.

At present, the fabrication of superconducting cables is attributed with grave technological hardships due to the brittleness and poor heat-conductivity of original materials, most often they are intermetalloids – chemical compounds of metals. This drawback is now overcome by producing superconducting multicore wires. The wire is manufactured (by pressing and drawing) from tin bronze reinforced by niobium filaments. When heated, the tin diffuses from bronze and forms an adsorptive superconducting film of niobium stannide Nb_3Sn on the filaments. The wires are twisted together into a multicore wire. When it is bended, the brittle superconducting film remains safe because this smart system is provided by FBS creating compressive stresses in the bronze matrix preventing the destruction of the stannide layer (taxon $\mathbf{A_1B_1}$).

One of the main applications of superconductors relates to creating super strong magnetic fields. The *superconducting solenoids* generate homogenous magnetic field with the intensity over 10^7 A/m, meanwhile the limit for the usual electromagnets with iron cores is the intensity about 10^5 A/m. The superconducting magnetic systems are provided with the FBS maintaining the circulation of non-attenuating current, therefore, they do not need any external power source. The superconducting solenoids reduce the overall size and energy consumption considerably in synchrophasotron and other amplifiers of elementary particles. They are used for plasma confinement in controlled thermonuclear synthesis reactors, in magnetohydrodynamic (MHD) converters of thermal energy into electrical energy, in inductive energy accumulators, as well as in systems of power peaks leveling in large power systems (Varivodov 2008).

To emphasize the availability and significance of smart superconductive devices, let us add the following data to the shown instances.

The application of superconductors in excitation coils of electrical machinery permits to obviate the cores from electrotechnical steel reducing the weight and overall size of machinery to 5÷7 times while saving the power. It is economically justifiable to develop the superconducting transformers with the power of the order 10÷100 MW. The pilot specimens of the pulsing superconducting coils have been developed to feed plasma guns and pumping systems of solid-body lasers. The superconducting volume resonators have begun to enter the radioengineering possessing very high robustness owing to negligible electrical resistance. The principle of mechanical repulsion of superconductors from the magnetic field is the basis of the new design of superfast railway transport on 'magnetic cushion'.

The high temperature superconductors (HTSC) of the second generation appeared in 2002 and intensified the application of the superconductivity technologies. Since 2004 the commercial projects of superconducting cables from HTSC have been underway. The company Sumitomo Electric completed the testing of the three-core superconducting cable for voltage $U = 66$ kV and nominal current $I = 1$ kA. Another cable ($U = 35$ kV, $I = 800$ A) was put into operation in the USA laid on the Hudson River bottom. The National Technological Center of power engineering and the Research Institute of cable industry (Russia) have prepared for testing the cable from HTSC with $U = 20$ kV and $I = 1500$ A. Experts predict that mass application of superconducting cables will begin in 2013÷2015 (Varivodov 2008).

A_5B_{19} is the class of conducting SM which, under the effect of electric field, generate the current adjusted by the FBS which changes the material conductivity type.

The film technologies are the techniques of manufacturing passive electro- and radioelements by applying on the dielectric substrate (board) conductive, resistive or dielectric layers. This technology is used to fabricate printed boards and integral microcircuits. Fine films up to fragments of nanometers (monoatomic layer) to several micrometers thick add to the passive elements attributes of activity because the volume of their surface layers is commensurable with the volume of the main material phase from which the film is formed. The forces of surface tension, which in the passive thick films are inessential, virtually determine the properties of thin films. The effect of these forces fulfilling in thin films the FBS functions is like the application of external pressure to the film which can alter the melting temperature of the film and the interplanar spacing in the crystalline lattice.

Thin metallic films are used in microelectronics as contact layers between elements of electrical circuits, capacitor plates, magnetic and resistive elements of microelectronic integrated circuits (IC).

The structure of the films formed by condensation in deep vacuum can change cardinally depending on the mode of condensation: an extremely disordered agglomerate of amorphous particles or perfect monocrystalline epitaxial layer can be obtained. In the films, the thickness of which is comparable with the electrons free path, the dimensional effects appear because, at the growing role of surface phenomena, the assumption of the independent specific material resistance on the geometric specimen dimensions becomes false. Fig 2.17 shows the dependence of specific resistance r on the temperature resistance coefficient (TRC) $\alpha = \dfrac{1}{R} \cdot \dfrac{dR}{dT}$ on the thickness of films.

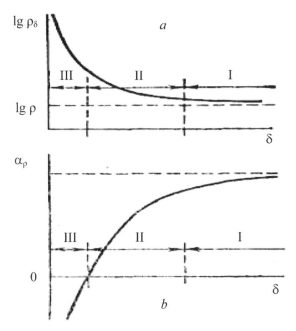

Fig 2.17. Typical dependence of specific resistance (*a*) and TRC (*b*) on metal films thickness. Dash lines correspond to values r and a of film metal

Three regions are singled out on curves *a* and *b*. Region I corresponds to films with thickness $\delta \geq 0.1$ μm, the specific resistance and TRC are close to the parameters of the original metal. In region II ($0.01 < \delta < 0.1$ μm), $\alpha \to 0$. Region III corresponds to films with thickness $\delta \approx 10^{-3}$ μm characterized by very high values of ρ and $\alpha < 0$.

Region III corresponds to the early stage of condensation when the films have the island structure. In the external electrical field at temperature rise, more and more electrons pass into the conducting band following the mechanisms of tunnel effect and thermal electron emission;

correspondingly the concentration of holes augments (Lazarev et al 1978). The islands in region II coalesce into conducting chains, and the TRC changes the sign. Then a continuous metallic layer appears on the substrate (region I). Its specific resistance is higher than that of the original metal due to a high concentration of vacancies, dislocations and boundaries of the grains formed when the islands coalesce. The impurity atoms migrate over the boundaries following the diffusion mechanism, that is why the even continuous thin films are not electrically continuous (Aleshin et al 1982).

The films with thickness $\delta = 3 \div 5$ nm (so-called Newtonian black films) have the structure corresponding to the structure of smart biological membranes (Pertsov 1995). It creates favorable prerequisites to supply the thin metallic films with the attributes of artificial intellect with the FBS adjusting reversibly the electrical conductivity of the island structure by the mechanism of surface diffusion. Now it is hard to imagine the physical and chemical mechanism of action of this FBS, but it is evident that its implementation would produce a revolution in the technology of semiconductor devices, membrane technology, IC technology, etc.

2.4.2. Semiconductors

Semiconductors are the substances which electrical conductivity at the room temperature has intermediate value between the electrical conductivity of metals ($10^6 \div 10^4$ Ohm$^{-1} \times$ cm^{-1}) and dielectrics ($10^{-10} \div 10^{-12}$ Ohm$^{-1} \times$ cm^{-1}). The concentration of movable charge carriers in semiconductors is considerably lower than the concentration of atoms and can vary in response to the temperature, illumination or a small quantity of impurities. These and the increasing with temperature conductivity make the semiconductors qualitatively different from metals.

In the intrinsic semiconductors, the states filled with electrons (energy levels) are separated from the vacant states by a forbidden band in which there are no electron states. Impurities and structural defects lead to the appearance of electron states (donors and acceptors) in the forbidden band (Fig 2.18), but the number of these states is few, so the notion of forbidden band retains its sense. The forbidden band width is the characteristic of the semiconductor and governs its electronic properties considerably. The adjustment of energy states of charge carried can be achieved with the FBS of temperature, emissive, electrical, magnetic and diffusive nature; it is the main methodological trend of the progress in smart semiconductors. The FBS can be implemented in semiconductor systems due to the unique properties of contact phenomena and photoconductivity.

Contact phenomena in semiconductors are unbalanced electronic processes which arise when electrical current passes the interface between semiconductor with metal or other semiconductor (hetero-transition) or

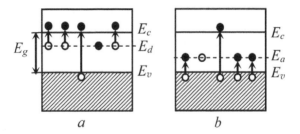

Fig 2.18. Band structure (energy diagram) of electron (*a*) and hole (*b*) semiconductors: E_c – bottom of conductivity band; E_v – ceiling of valence band; E_g – gap energy (width of forbidden band); E_d and E_a – energy levels of donors and acceptors; dark circles correspond to electrons, light circles – to holes

across the boundary between two regions of the same semiconductor (homo-transition) differing by the types of charge carriers (electron-hole or *p-n*-junction) or their various concentration. The contacts of semiconductor with metal or another semiconductor pass much more effectively the electrical current in one direction than in the other. The dependence of the current through *p-n*-junction from the applied voltage (the volt-ampere characteristic) has a strongly pronounced non-linearity (Fig 2.19). The value of the current through the *p-n*-junction changes $10^5 \div 10^6$ times when U reverses the sign. Due to this, the electron-hole junction effect is similar to the action of the gate unit and is used in smart devices of the converting engineering, such as transformers, rectifiers, inventors, frequency converters, phase splitters, etc. Semiconductor converters assuring contactless commutation of currents in power circuits are characterized by heightened reliability and energy indicators, smaller overall sizes and weight.

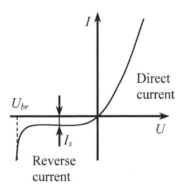

Fig 2.19. Volt-ampere characteristic of *p-n*- junction: I_s – saturation current; U_p– puncture potential

$A_5B_{17,19}$ are smart semiconductor converters reacting to the effect of electrical field by the onset of current the value of which is adjusted by FBS correcting the conductivity type in the interface between the semiconductor and the conjugated element.

The semiconductor diode is the element of electrical circuit with one-way conductivity depending on the polarity of applied voltage. The diodes action is based on the properties of *p-n-* junction or transition metal-semiconductor. The FBS in diodes is fulfilled by the following (Gergel 1988):

- by diodes from highly alloyed (degenerated) semiconductors – the transparency of the energy barrier between *p-* and *n-* regions making for tunneling of electrons from valence band into the conductivity band and backwards;
- by avalanche diodes generating SHF emission – a special distribution profile of alloying elements in semiconductor making for neighbourhood of the narrow region with high field intensity (the region of avalanche multiplication of carriers) and the region with weak field (the drift region or the passage region);
- by diodes amplifying the SHF emission – the appearance of volume negative resistance due to particular distribution of electrons in semiconductor (the Gunn effect).

Thyristor is a semiconductor device consisting of three *p-n-* junctions. Their interaction results in one of two stable states of the device: the switched off state – with high resistance, and switched on state – with low resistance. Transition of the smart device into the on state takes place when the applied voltage increases adjusted by the FBS initiating the growth of the current injecting component of the emitter transitions and the field acceleration of carriers transition through internal *n*-layer. This process grows snowballing and results in filling up the internal *p-* and *n*-layers with electron-hole gas of large density. The thyristor is switched off by short-time change of external voltage polarity. The current changes the direction, and the FBS extracting the charge carriers from gas into external circuit is switching on (Tuchkevich and Grekhov. 1988).

The main original semiconductor materials for fabricating the thyristor are silicon and gallium arsenide. The laminar *p-n*-structure is produced by thermodiffusion of *p-* and *n*-types impurities into the monocrystalline semiconductor wafer.

Symistor (symmetric thyristor) adjusts the current flow in both directions. The semiconducting structure of the symistor contains five layers semiconductors and has a more complicated configuration compared with the thyristor. The combination of *p-* and *n*-layers creates the semiconducting structure in which the FBS is acting determining the rectilinear volt-ampere characteristic of the symistor at different voltage polarity.

$A_3B_{17,19}$ are smart semiconducting measuring converters of temperature in which the current is controlled by the FBS adjusting the voltage at the semiconductor or the conductivity of *p-n*-junction.

Thermistor (thermoresistor) is a semiconducting device adjusting the temperature based on the dependence $I = \sigma(T)\cdot E$, where I is the current , σ – semiconductor electrical conductivity, T – temperature, E – external field intensity. The thyristors have small (on the order of millimeters) dimensions and long (thousand hours) service life. The spheres of application are the systems of heat control, power meters, magnetometers, etc. The FBS function in thyristor is executed by external automatic controller of intensity E.

The direct current is registered at the thyristor output, what is not quite convenient for signal amplification. That is why the circuits of converters are used in a state of two diodes or a semiconducting monocrystal with two *p-n*-junctions receiving the output signal in the alternating voltage form.

Two serially connected measuring bridges with diodes placed into the bridges diagonals serve to measure the temperature difference. The alternating current is registered equal to the difference of currents in bridge circuits. Its amplitude is the function of temperature difference in the points of the bridges placement (Pasynkov et al 1987).

Thermal emission converter (TEC) is a device for direct conversion of plasma thermal energy into electricity. The classic TEC type is a diode to the cathode of which thermal energy is applied, and the electrical energy is gathered in loading resistance. In such a way the phenomenon of thermal ion emission (the substance release of ions when heated) discovered by Edison in 1883 is realized. In the semiconducting TEC type developed by the physicists of the Massachusetts Institute of Technological, the role of vacuum clearance between diode electrodes is played by the surface layer of the semiconducting wafer from indium antimonide enriched by charge carriers. This makes it possible to intensify considerably the thermal ion current at modest temperatures (~200 °C). The nature of the observed effect still remains unclear. It is assumed that this converter has the FBS initiating the chain reaction of current snowballing amplification in the electron-enriched semiconducting layer between the TEC emitter and collector.

The photoconductivity (PC), or photoresistive effect, is the rise of the semiconductor electrical conductivity under the effect of electromagnetic emission. For the first time the PC induced by concentration increase of moveable charge carriers under light action (the concentration PC) was observed in selenium by W. Smith (USA) in 1873.

The main causes of PC initiation are the following (Fig 2.20): photons extracts electrons from the valence band and 'throw' them into the conduction band, so the number of conduction electrons and holes grows simultaneously (the intrinsic PC); the electrons from the filled band are 'thrown' onto the free impurity levels enlarging the number of holes (the hole impurity PC); the electrons are 'thrown' from impurity levels into the conducting band (the electron impurity PC).

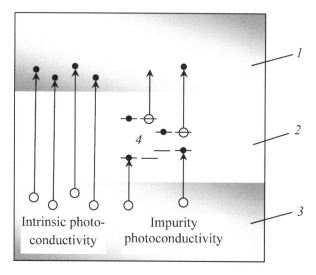

Fig 2.20. Formation of different types of photoconductivity in semiconductors: *1* –conductivity band, *2* – forbidden band, *3* – valence band, *4* – impurity levels

The concentration PC can appear only due to excitation by sufficiently short-wave emission when the energy of photons exceeds the energy gap or the distance between one of the bands and the impurity level. The concentration PC value is proportional to the quantum yield Y (the ratio between the number of produced carriers and the total number of absorbed photons) and the life time of non-equilibrium carriers excited by light. When illuminated with visible light, $Y < 1$ owing to the competitive processes of light absorption is non-connected with carriers formation (excitation of excitons, impurity atoms, phonons, etc.). When the substances are irradiated with UV or more hard radiation, $Y > 1$ because the photon energy is be sufficient not only to extort the electrons from the filled band but also to impart it the kinetic energy by the mechanism of impact ionization. The carrier life-time (the time t which it spends in free state) is determined by the kinetics of recombination processes. At direct (between bands) recombination, the electron moves at once from the conductivity band into the valence band. In case of recombination through impurity centers, the electron is first trapped by the impurity center and only after that it gets into the valence band. The carrier life-time ($1 \text{ s} > t > 10^{-8}$ s) depends on the material structure, its purity and temperature. The idea of creating smart PC systems consists in optimization of charge carrier's separation in the semiconductor using the FBS.

A_4B_{17} are smart semiconducting devices transforming light energy into electrical energy under control of FBS adjusting the separation of charge carriers at the boundary of the *p-n*-junction.

Photodiode is a semiconductor diode with one-way concentration PC appearing under the influence of optic emission. It operates in the following manner. When the light is falling normally to the direction of *p–n*-junction, the electron-hole pairs appear in the *n*-region as a result of absorption of photons with energy exceeding the energy gap. When they diffuse deep into the *n*-region, the main part of electrons and holes have no time to recombine and they reach the boundary of *p–n*-junction. Here the pairs are divided by the electric field of *p–n*-junction which fulfills the FBS function. The holes pass into the *p*-region while the electrons can't overcome the transition field and accumulate near the *n*-region boundary. Thereby, the current through *p–n*-junction (photocurrent) is conditioned on the drift of minority carriers – holes. They charge the *p*-region positive relative to *n*-region, while electrons charge *n*-region negative relative to *p*-region. As a result, potential (photo-voltage) appears on the *p–n*-junction boundary. The current generated in the photodiode is the reverse one, in other words, it is directed from cathode to anode, and as its value is the larger the more intensive is the illumination.

The photodiode nanotubes are components of light-absorbing diodes and transistors developed in Japan (Yamamoto et al 2006). Nanotubes (16 nm in diameter, several micrometers long) consist of the outer electron accepting and inner electron-donor layers. The light conductivity of nanotubes is 10^4 times higher than in darkness dictating a large difference between the photocurrents when operating in 'on – off' regime. The FBS in these smart photoelements acts similarly to photodiodes.

The photodiodes are used in the means of automatics, computer and measuring equipment as generators of photo-EMF or elements of current control in electrical circuits.

A_5B_{16} are smart semiconducting photodevices generating electromagnetic waves in the electrical field under control of FBS adjusting the recombination of charge carriers in semiconductor.

The light emitting diode (LED) is a semiconducting diode emitting light when the current passes through *p–n*-junction or hetero-transition in direct direction. It causes the recombination of electrons and holes and emission of photons due to the electrons transition to a lower energy level. The spectral characteristics of emitted light depend on the chemical nature of semiconductors. The best light emitting diodes contain the so-called direct band semiconductors in which the direct optic transitions are permitted between bands. They are chemical compounds of the type $A^{III}B^V$ (for instance, GaAs or InP) and $A^{II}B^{VI}$ (for instance, ZnSe or CdTe).

Variation of the composition of semiconductors permits one to create light emitting diodes operating in the wavelength band from ultraviolet (GaN) to average infrared band (PbS). It is possible to imagine the smart light emitting diodes which spectral characteristics are adjusted by the FBS initiating new optic transitions, but not by varying the chemical composition of semiconductors.

The light emitting diodes are characterized by high brightness (thousands cd/m^2), light intensity (up to ten cd), emission force (hundreds mW/sr), speed of response (ns), compactness and weight (the overall sizes of active element from semiconducting crystal are 0.3×0.3×0.25 mm) (Kogan et al 1990). Their input characteristics make them compatible with transistor IC, the emission spectrum makes them compatible with photodiodes.

The spheres of light diodes application are: signal indication, illumination of permanent inscriptions, imaging of mnemonic information (facilitating memorizing), systems of running lines and large displays, remote control devices in household and industrial radioequipment, illumination of copying and reading devices of personal computers, image analyzers, devices for contactless measurement of angles and angular displacements, etc.

Optron is a semiconducting device consisting of light emitting diode and photodiode built in common body and associated by optical and electrical links. The photodiode light sensitive plate is placed opposite the emitting one of the light emitting diode (Fig 2.21).

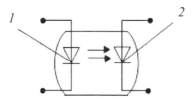

Fig 2.21. Optron diagram: *1* – light emitting diode, *2* – photodiode

The optrons main application is to provide the galvanic decoupling between signal circuits. The principle of action of these devices is that the input electrical signal entered the emitter is converted into the light flux which affecting the photoreceiver changes its conductivity. The optron generates at the output a signal which can differ by shape from the one at the input so that the input and output circuits are not connected galvanically. A transparent dielectric spacer (usually, from organic polymer) with the resistance $10^9 \div 10^{12}$ Ohm is placed between the optron input and output circuits. Thus the optrons can be referred to the smart systems in which two types of the FBS are implemented: the electrical ($\mathbf{A_4B_{17}}$) and the emitting one($\mathbf{A_5B_{16}}$).

The optrons have found broad application in circuits of matching microelectronic logical units with executive devices (relays, electrical motors, contactors , etc.), and as a coupling element between logical units needing galvanic decoupling, as modulators of constant and slowly changing voltages, converters of rectangular pulses into sinusoidal oscillations, control of powerful lamps and high voltage indicators.

The electro-, radio-, light-, heat- and computing equipment, as well as management systems of industrial articles and technological processes in all branches of production are now hardly imaginable without smart semiconducting materials and devices.

2.4.3. The dielectrics

The dielectrics are substances poorly conducting electrical current. The active dielectrics are most interesting as the SM basis. Their properties depend considerably on external conditions, such as temperature, pressure, field intensity, etc. These dielectrics act as working bodies in sensors, converters, generators, modulators, other active elements in electroengineering and radioelectronic equipment.

The active dielectrics are ferro-, pyro- and piezoelectrics, electrets and materials of quantum electronics. One and the same material may display the attributes of active dielectrics or, depending on the operation conditions, can act as passive insulator or active converting and controlling element.

The usual (passive) dielectrics are not polarized or there is no external electrical field because the existing internal electrical field is compensated by the field of free charges flowing to specimen surface from inside and outside. Violation of this compensation leading to polarization (temporary appearance of electric field) occurs in piezoelectrics (at certain deformations) and in pyroelectrics (at temperature changes). The variety of pyroelectrics are ferroelectrics which spontaneous polarization can change sufficiently under external effects. The induced by external field polarization remains long in electrets (polymeric and ceramic) after the polarizing field is removed.

In principle, the active dielectrics are multifunctional or adaptive materials. However, the external FBS can attach the properties of smart system to the article or device. Many active dielectrics possess internal (or natural) FBS which also enables them to implement the attributes of artificial intellect in technical systems.

Publications refer most often to the SM piezoelectrics (alpha-quarz, lead titanate-zyrconate). They act as sensors or actuators in high-precision pointing devices, particularly, in scanning probe microscopy (Mironov 2004). Piezoelectrics have served to develop a number of so-called nanodevices: nano- and microscales, fine-film nanostructures from barium titanate or zinc oxide used to generate electricity, etc. (Suzdalev 2009).

The direct piezoelectric effect is used in pressure sensors in which the piezoelectric element generates an electrical signal proportional to the acting force or pressure. Piezoelectric sensors with broad dynamic and frequency bands have a little weight and outer dimensions, high reliability and serve to measure fast-variable acoustic and mechanical pressures in rigid operating conditions (Sharapov et al 2006). This system can be referred to taxon A_1B_{17} if there is a FBS of electrical nature adjusting the piezoelement deformation value under the effect of external forces. Let us show examples of smart piezoelectric systems.

The piezoelectric generator assigned to receive electricity from the highway bedding was developed by Israeli Innowattech Co. The energy source is the pressure which exerts on the setting the moving vehicle, train or aircraft during take-off or landing. The advantages of this business-idea versus other developments of environment friendly energy generation is that its implementation does not need extra area, the environment is not damaged, the system operates irrespective of weather conditions. Technically it is realized in the following way. The piezoelectric generators which convert the pressure energy of the moving transport into electric energy are spaced-apart and placed under the motorway asphalt, the take-off runway or the rails. The energy is saved in compact accumulators and delivered directly to consumers.

The piezogeneration of electricity has been implemented in significant public consumption scale. The Japanese metro has a station with piezoelectrical flooring where the pressure of passenger feet serves to generate power sufficient for several turnstiles feeding. The dancing floor in a British disco club is equipped with piezoelectrical sensors converting the impact energy of dancers into electricity. The more the dance is intensive, the brighter the light-show lights. The Innowattech Co. installed the piezogenerators at the pilot railway leg. The developers assert that the traffic of 10÷20 trains per hour is sufficient to provide overall with electricity to 150 lived in houses. The smart system is the most acceptable characteristic of these generators in spite of the fact that they have no FBS. In principle the FBS is not necessary, because the more electric energy will be produced the better. FBS in such systems will be expedient only for decreasing the fatigue damage of piezogenerators.

The inverse piezoelectric effect is implemented in many SM and smart systems. *The smart bearing design* in which vibration reduces friction is provided with the deformation FBS (taxon A_2B_3) (Broeze and Laubendorfer 1966). The bearing sleeves are made from piezoelectrical material provided from both sides with foil electrodes which are fed with alternating voltage. The current flowing through electrodes 'enforces' the piezoelectrical sleeves to shrink and expand in response to the applied voltage inducing vibration which reduces the friction like lubrication.

A_4B_3 and A_5B_3 correspond to the class of SM performing mechanical work in response to the emission or electrical current with the deformation FBS adjusting the input signal.

Piezoelectrical resonator (PR) is a piezoelectrical converter with pronounced resonant properties close to the own oscillation frequencies. The PR usually consists of a fine piezoelectrical plate sandwiched between two electrodes. The plate is of quartz crystal or piezoceramics with low losses for hysteresis (Vysotskii and Dmitriev 1985). The radiofrequency signal coming to the plate excites in the latter mechanical oscillations which are in resonance with its own oscillations. To this end, the plate thickness should be half of the input signal wavelength. If the frequencies of oscillations coincide, the FBS switches off the input signal automatically fixing its frequency. PRs are applied in navigating systems for frequency accurate checking, in the systems of communication with satellites, etc.

The micro-miniature quartz PR operating at the frequency of oscillations from 30 kHz to 8.4 MHz has found broad application in radioengineering, electronics, electroacoustics. They serve as filters, resonators in driving oscillators, resonance piezoconverters and piezoelectric transformers, as well as in electronic clocks, systems of electronic ignition in internal combustion engines and other smart devices. It is promising to apply the PR as sensors to register variation of specimen mass, for instance, after gases adsorption (Suzdalev 2009).

The electret state is peculiar to dielectrics possessing quasi-constant electrical charge (Sessler and West 1987). The term 'quasi-constant' means the time characterizing the electret discharge which is much longer than the time during which the electret is studied. The electrets serve as sources of direct electrical fields in converting devices (electret microphones and telephones, vibration detectors, electrometers, electrostatic generators), as well as in gas filters, barometers, dosimeters, etc. (Gubkin 1978, Sessler and West 1987).

A_1B_{17} – class of smart technical systems in which, due to the internal electret deformative FBS, the acoustic or mechanical oscillations are converted into electrical signals.

The electret microphone (Fig 2.22) is a typical example of the acoustic converter. The membranes from polymeric film *1* with charge surface density $10 \div 20$ nC/cm^2 has the metallized upper surface. It is arranged over the stationary electrode *5* which is provided with microlugs to maintain the optimal clearance (10–30 μm) between membrane and electrode. Under the effect of sound wave, the membrane vibrates and alters the potentials difference between metallic coating on the membrane and stationary electrode *5*. This signal is transmitted to the telephone and again converted into the acoustic signal. Apertures *6* in electrode allow to reduce the resonance frequency of the output signal and to amplify the microphone sensitivity. When the electret membranes are used, there is no need of external voltage sours which is usually used by condensing microphone. The outer signal converter is not required, that is why the preamplifier device is simplified significantly. As a total result, the dimensions, weight and cost of the electret microphone are considerably less than those of the condensing microphone.

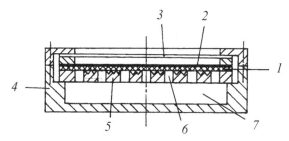

Fig 2.22. Diagram of electret microphone: *1*– electret polymeric film, *2* – metallic coating, *3* – safety fabric, *4* – body, *5* – stationary electrode, *6* – apertures, *7* – air chamber

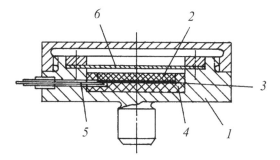

Fig 2.23. Diagram of measuring vibration converter: *1* – body, *2* – electret, *3* – metal coating, *4* – dielectric spacer, *5* – cable , *6* – membrane

The electret vibration measuring converter has the design shown in Fig 2.23. The electret plate 2 with lower metallized surface is fixed in the body with dielectric spacer 4. The converter is mounted on the machine vibrating part to vibrate together. Then membrane 6 oscillates with respect to the body and rigidly connected with it electret plate. The inductive current appears in the external circuit and its frequency is equal to that of the machine vibration. The converters of pressure, displacements, deformation are designed similarly.

A_5B_{17} are smart systems in which the current appears charging the accumulative capacitor due to the electret internal FBS.

The electret power sources induce alternating current in the electret constant electrical field, or as a result of interaction between the fields of electret and electrodes. One of the first models of high voltage generators on electrets was proposed by A.N. Gubkin (Gubkin 1978). The generator consists of a flat electret element sandwiched between moveable and earthed stationary electrodes. When the contact between the moveable electrode and the electret is broken, the charges induced at the electrodes flow to the capacitor. The moveable electrode is earthed when it contacts with the electret, and switched off from the earth before the contact is broken. By multiple repetition of this operation, the capacitor is charged to the voltage equal to the potential difference between the surfaces of the electret element. The operating generator model charged a moderate capacity up to 2 kV during 1 min making the power about 0.1 mW. Further improvement of electrostatic generators design on the electrets base enabled to reach the current values up to 1 mA and the power output up to 1 W (Tada 1992, Merkulov and Golushko 2009).

2.4.4.   Magnetic materials

The specific of magnetic materials application in electrical and electronic engineering is determined by the total combination of their properties. In this connection, one can single out the next main groups of magnetic materials: magnetostrictional, thermomagnetic, magnetooptical and materials with giant magnetoresistance.

Magnetostrictional materials are utilized in ultrasonic generators and receivers, acoustical converters, magnetostrictional mechanisms of microdisplacement, delay lines of sound and electrical signals, and other devices of radioengineering and electrocommunication.

A_6B_3 is the class of SM converting the magnetic field energy into the mechanical (sound or ultrasonic) energy under the deformational FBS effect.

Magnetostrictional converters implementing the megnatoelastic effect are the basic desing of smart systems of this class. If the magnetoelastic rod is arranged along the variable magnetic field, it begins to shrink or elongate, that is to undergo the mechanical oscillations with the frequency of variable magnetic field and the amplitude proportional to its induction. The rod oscillations excite in the solid or liquid medium which the rod contacts the ultrasonic waves of the same frequency.

The main design types of magnetostrictional converters are the bar and the annular types (Fig 2.24). When alternating current flows through the winding carrying the core-strictor, oscillations appear in the strictor corresponding to the frequency of the generator electrical signal. The FBS functions in these devices are fulfilled by the property of magnetostrictional material consisting in the nonlinear deformation dependence on the magnetic effect force.

The metallic alloys with high strength characteristics serve to produce most powerful magnetostrictional converters, including those with giant magnetostriction (Belov 1998). But the high electrical conductivity of the alloys causes the remagnetization loss and considerable eddy-current loss reducing the device efficiency. To reduce the eddy currents, the converters are made of package of metal plates of 0.1÷0.2 mm thick. In spite of this, the converters from magnetic alloys demonstrate poor efficiency (40÷50 %) necessitating their cooling. The ferrite converters possess better efficiency (70%), because of high electrical resistance they are free of the Foucault current loss, but their power characteristics are considerably limited by poor mechanical strength.

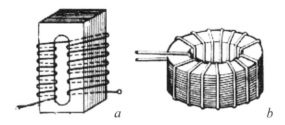

Fig 2.24. Bar (*a*) and annular (*b*) magneostrictional converters

The magnetostrictional converters are used in the ultrasonic defectoscopy as broad band sensors of oscillations, in acoustic electronics as filters and resonators, in ultrasonic cleaning processes and degreasing of workpieces.

The magnetostrictional floating-type sensor has the float as the main design element containing a permanent magnet (Fig 2.25). The float moves freely vertically sliding along the wave guide from magnetostrictional material. The electronic unit of the sensor generates current pulse with the set periodicity which propagates along the wave guide. When the pulse reaches the float, it magnetic field interacts with the float's magnetic field inducing mechanical oscillations which propagate back along the wave guide and are registered by the sensitive piezoelement. The time delay between the current pulse dispatch and the moment when mechanical pulse is received, enables one to rate the distance to the float, and so – the liquid level in the tank. Therefore, the deformation FBS induced by the interaction between magnetic fields adjusts the mechanical pulse in the magnetostrictional wave guide. The magnetostrictional sensors are very precise and can operate with the flexible wave guide extending the application sphere. One of such spheres of float sensor application are the devices for objects check deviation from vertical during orienting in the gravitational space (Vorontsov et al. 2010).

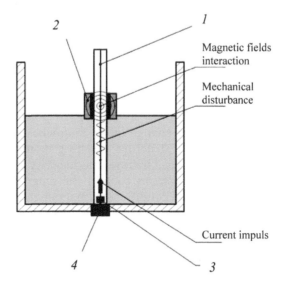

Fig 2.25. Diagram of magnetostrictional float sensor operation: *1* –magnetostrictional wave guide, *2* – float with magnet, *3* – piezosensor , *4* – current pulse source

Thermomagnetic materials are ferromagnetic alloys with sharp dependence of the magnetic saturation induction B_S (or magnetization) on the temperature.

A_3B_{20} are SM in which temperature variations alter the magnetization under the control of FBS restructuring the material domain structure. The sharp temperature dependence on magnetization of this class of SM is the consequence of ferromagnetic properties to reduce the magnetic induction as the temperature grows around the Curie point. The thermomagnetic materials are the alloys with Curie point $T_C = 0 \div 100$ °C (nickel alloy with copper, iron alloy with nickel). The broadest application got the Ni-Cr-Fe alloys in which the chrome content determines the Curie temperature from 70 to 120 °C.

Thermocompensators and thermoregulators of magnetic flux in measuring instruments (galvanometers, electricity meters, speed meters, etc.) are made as shunts taping to itself a part of the permanent magnetic flux. When the temperature changes, the shunt magnetization diminishes abruptly, hence, the magnetic flux in the magnet clearance jumps. Thanks to this fact, the instrumental error concerned with the temperature-induced changes of magnetic induction and electrical resistance of the measuring winding are compensated. Thermomagnetic alloys are also used in relays which operation time depends on the temperature (Bishard 1988).

A_6B_{10} are SM in which the variable magnetic field causes the phase transition of the second order under the control of temperature FBS

The magnetocalorical effect (MCE) is the temperature variation of the magnetic material induced by its magnetization or demagnetization in adiabatic conditions. The MCE reaches the maximum value at the temperature of magnetic phase transition, for instance, at the Curie temperature of ferromagnetics. The magnetic field application causes the ferromagnetic heating, and field removing – its cooling. In practice, the process adiabaticity is reached by fast change of the magnetic field. The FBS is built into the thermomagnetic structure adjusting its magnetization by nonlinear variation of the material thermal conductivity around T_C.

Magnetic refrigerating installations contain the magnetic material as the working body. The optimal conditions of magnetization and demagnetization permit to achieve a considerable temperature reduction. The magnetic working body acts as the coolant used in steam-to-gas refrigerating units, and the magnetization/demagnetization process serves as an analog of the compression/expansion cycle (Van Geuns 1968). The most promising materials for magnetic cooling are gadolinium and intermetallic compound of gadolinium silicide-germanide $Gd_5Ge_2Si_2$. The magnetocalorical effect in gadolinium reaches about 3 degrees when the magnetic field induction changes by the value 1 T.

The so-called active magnetic regenerative (AMR) cycle is applied in magnetic refrigerating installations. Fig 2.26 shows the typical diagram of the refrigerator working circuit (Tishin and Spichkin 2005). The AMR cycle consists of two adiabatic stages (magnetization/demagnetization) and two stages of heat-transfer agent blowing through the circuit carried out in the permanent magnetic field. At the first stage, piston 6 of blower 5 is at the rightmost position (the heat-transfer agent is in the cold heat exchanger 3), and the magnetic material in regenerator 2 is magnetized adiabatically which provokes its temperature rise by the MCE value. At the second cycle stage (hot blowing), the blower moves the heat-transfer agent from cold heat exchanger 3 to hot one 4. The heat released during magnetization in magnetic regenerator 2 is transferred to the heat-transfer agent and is liberated to environment in heat exchanger. At the third cycle stage, when the piston is at the leftmost position and the heat-transfer agent does not move in the circuit, the magnetic material in regenerator 2 is demagnetized adiabatically which provokes its cooling by the MCE value. At the fourth finishing cycle stage (cold blowing), the heat-transfer agent moves, under the effect of the blower pump, in the reverse direction (from the hot heat exchanger 4 to the cold one 3), cools in the regenerator and is pumped into the heat exchanger where it cools the loading. Multiple cycle repetition cools the cold heat exchanger 3, because the heat is removed from the loading and is released into environment in the hot heat exchanger 4. The FBS function in the smart refrigerator is performed by the system of magnetization/demagnetization of the regenerator and hot/cold blowing of the heat-transfer agent through the circuit.

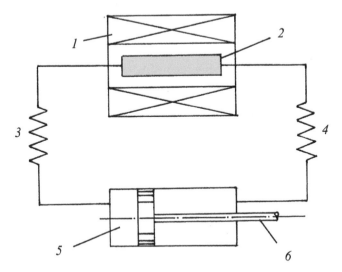

Fig 2.26. Typical diagram of working circuit of AMR refrigerator: *1* – magnet; *2* – active magnetic regenerator; *3* and *4* – cold and hot heat exchangers; *5* – reversible heat-transfer agent blower; *6* – piston

The treatment of malignant growth by MCE application needs special medical equipment (Tishin 2007). The growth with pre-injected magnetic fluid having high MCE and phase transition temperature close to that of human body is exposed to variable magnetic field. Its amplitude and frequency are selected by the criterion of safe MCE value with the account of particles concentration and malignancy dimensions. Under the effect of the magnetic field, the temperature of particles grows, and they reverberate the released heat to the surrounding tissues. The particles are magnetized periodically until the malignancy temperature reaches 43 °C when the affected cells perish.

The optical activity induced by the magnetic field in *magnetooptical materials* appears in the Faraday and Kerr effects. The Faraday effect consists in rotation of the linear polarization plane of the light beam passing through the substance placed into the magnetic field. The angle of rotation is proportional to the magnetic field strength when the magnetic field is directed along the beam. The Kerr effect is observed in the magnetic field when the linearly polarized light beam is reflected by the surface of ferromagnetic specimen. Thus, the passed or reflected light carries the information about the field intensity in ferromagnetic material registered by the angle of rotation of beam polarization plane. The beam modulation by polarization is converted into the modulation of light intensity using the system of crossed polarizers. The modulated light intensity is proportional to the squared magnetic-field strength.

A_4B_{21} is the SM class in which the ferromagnetic undergoes the domain restructuring under the effect of optical emission with the magnetic FBS participation.

Magnetooptical disks are used to record and read information. The principle of magnetoptical (MO) recording is based on the application of magnetic materials requiring heating over the Curie point for remagnetization. A typical MO-disk (Fig 2.27) is a polycarbonate substrate 1.2 mm thick coated with several thin functional layers, including the magnetic one. The structure of the magnetic layer of ferromagnetic alloy TbFeCo (terbium-iron-cobalt) consists of domains oriented perpendicularly to the disk surface.

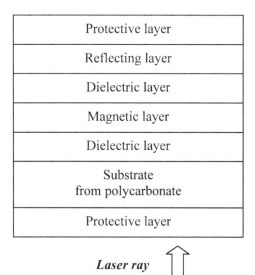

Fig 2.27. Structure of MO-disk

The disk preparation to introducing information consists of orientation of all domains in one direction. When heated to the Curie temperature ($T_C \approx 200$ °C for TbFeCo), the magnetic disordering occurs: the ferromagnetic loses the domain structure and becomes demagnetized. The domains get oriented in the external magnetic field direction if the field is applied during heating (Fig 2.28, *a*).

The information is recorded on thus prepared disk (*b*). At the time of the head passing, the necessary bytes are transferred from the original

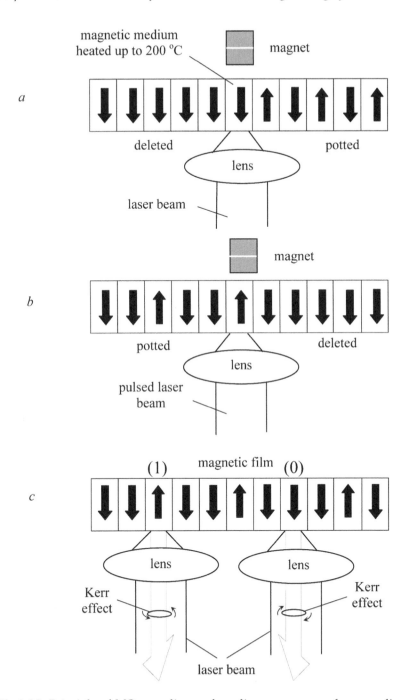

Fig 2.28. Principle of MO-recording and reading: *a* – erasure; *b* – recording; *c* – reading

'zero' state into the 'one' state under the head field effect. The data reading (*c*) is realized with the help of polarized laser beam of reduced power (25% of the nominal) which is insufficient to heat the disk magnetic layer. When the beam hits the disordered at the time of recording disk magnetic portions, their magnetic field changes the beam polarization (the Kerr effect). Though the polarization plane deviates just for several degrees, it is enough to record and read information.

In spite of technical perfection (in line with the present day know-how), the system of MO processing of information is rather actively-adaptive than a smart system. It is expedient to provide it by FBS which protects the disk from ageing at multiple overwriting and overheating in operation and storage in unfavorable conditions.

Giant magnetoresistance (GMR) is the effect of increasing by 12 percent the specimen electrical resistance under the effect of magnetic field, unlike that of the usual magnetoresistance when the resistance change does not exceed several percent. The GMR was discovered in 1988–89 by two independent groups of researchers led by A. Fert and P. Grunberg; they were awarded the Nobel prize in physics in 2007 (Nobel laureates 2007). The GMR is observed in multilayered materials with alternating ferromagnetic and non-magnetic metallic thin layers (Baklitskaya 2007).

A_6B_{20} are SM in which, under the effect of magnetic field, the electrical resistance grows under controlling by the magnetic FBS.

Multilayered materials demonstrating the magnetoresistance effect consist of alternating layers of conductors and ferromagnetics. The layers are from fractions to units of nanometers thick. The resistance to the current in conducting layer depends on the magnetization direction of adjacent magnetic layers. In case of parallel magnetization, the layers as a rule have low resistance, while at non-parallel it is high (Fig 2.29). The resistance of multilayered structures changes from 5 to 50 % of the nominal resistance depending on materials nature, number of layers and temperature.

These materials are used over a decade as magneto-sensitive elements in various sensors, particularly in the reading heads of hard disks. The mass production of magnetoresistive elements of the so-called magnetic random access memory (MRAM) began (Wolf et al. 2001). The MRAM system excels by a number of indicators the semiconducting counterparts.

2.4.5.   Optically active materials

Strictly speaking, the optical activity is the substance property to induce rotation of the plane of polarization when the plane-polarized light passes through the substance. The natural optical activity is inherent to protein, enzymes, vitamins, other natural compounds. This paragraph deals with

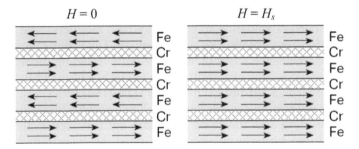

Fig 2.29. Alternation of conducting (Cr) and ferromagnetic (Fe) layers with antiparallel (*a*) and parallel (*b*) direction of magnetization controlled by external magnetic field *H*

the man-created materials and articles through which optical activity is understood more broadly.

The notion 'optically active' incorporates objects of three classes: i) those having optical characteristics which vary due to external effects: temperature, electromagnetic fields, sound, etc. (the brief term is 'dissipating and anisotropic'); ii) the physical and chemical parameters of which are controlled by the electromagnetic waves of optical band – $10^{13} \div 10^{15}$ Hz ('photofunctional'); iii) light sources. Almost all instances below correspond to the taxons $\mathbf{A_4 B_{13,16}}$: their reaction to illumination is corrected by the FBS purposefully restructuring the SM or modifying the kinetics of particles and radiation emission. Several other taxons are particularly characterized.

(i) **Dissipating and anisotropic materials** can demonstrate three principally different optical effects. The dissipating materials deflect the light beam in all possible directions. The limiting states of anisotropic materials are transparent or have the property to pass directed emission, and opacity, or full light absorption by the material.

The optical gate is the device to shut off or pass the light beam according to the assigned program. The *electrooptical gate* function is based on the linear (the Pockels effect) or quadratic (the Kerr effect) electrooptical effects – the dependence of medium double refraction on the applied electrical field intensity. The FBS function in this gate is performed by the property of active optical SM to ensure the phase shift between normal and specific waves which magnitude reaches p when the so-called 'half-wave voltage' is applied inducing double light refraction.

The action of *magnetooptical gate* is based on the linear magnetooptical effect (the Faraday effect) – dependence of Faraday rotation angle of the light in medium on the applied magnetic field intensity.

The FBS function in this instrument is performed by the smart system of current adjusting in solenoid which magnetic field controls the transparency of the gate active element. The FBS of magnetooptical gate creates the effect of 'non-mutuality' opening the light passing in one direction and shutting off the light passing in the reverse direction (so-called 'optical insulator') (Zhdanov 1992).

The polarizing devices are optical instruments meant for receiving, detecting and analyzing polarized light, as well as the instruments in which light polarization is used for measurements. They are based on the light interaction with the optical active material disturbing the light beam axial symmetry. Many instruments use the optical bistability – the effect of light self-adjustment when it passes through optical active SM with nonlinear polarization. The self-adjustment is controlled by the FBS converting the incident emission into two stable stationary states of the passed wave. They have different amplitude and/or polarization parameters as well as pronounced hysteresis properties. The principle of action of optical computers and light guides is based on the optical bistability.

The optical computers are the generation of computing equipment in which the optical emission serves as information carrier. They possess excellent performance (the light signal propagates faster than the electrical one), they are able to combine parallel data processing (the interaction between light fields with the linear optical active element is not localized in space but distributed over the entire volume). Due to this, the digital optical processor with the number of parallel channels $10^5 \div 10^6$ is able to perform up to $10^{13} \div 10^{15}$ operations per second (the channel commutation time is $10^{-8} \div 10^{-9}$ s) (Sinitsyn 1987). It saves the energy losses by obviating multiple conversions of electric energy into light and back. The logical devices of optical computers possess the rigid positive FBS under the control of which the element 'YES' is on state if the element 'NO' is off, and vice versa. The optical computers are a new generation of computing equipment of which parameters are improved continuously as the special elementary base progresses.

The optical fiber is a device for direct transmission of light energy. It is a thin filament of which the core has a refractive index larger than that of the shell. A considerable contribution to the light propagation along the fiber is made by the total internal reflection from the boundary core-shell. The optical bistable glass fibers demonstrate the optical non-linearity, in other words, the dependence of refractive index on the emission intensity. The FBS optimizes the non-linearity parameters. That's why, when the laser emission with the power 1 kW is introduced in the smart fiber, very high emission intensity (~ 1 MW/cm^2) remains throughout the length of several kilometers (Dianov and Prokhorov 1990).

The elements of the instruments with the optical instability are miniature, with low energy consumption and short commutation time (~ 10^{-12} s).

The photochromic materials implement the phenomenon of photochromism – the ability, under the effect of light, reversibly alter the absorption spectrum in the visible band what appears in color changing of the original material or coloring appearance of the earlier transparent material. The photochromical materials are used in the states of liquid solutions, polymer films, amorphous or polycrystalline thin coatings applied on flexible or rigid substrates, silicate or polymeric glasses, single crystals. The FBS determines the reversibility of photophysical and photochemical processes in photochromic SM.

The photochromic silicate glasses containing microcrystals of silver halogenides (AgBr, AgCl) demonstrate virtually unlimited cyclicity of the process 'photo-induced coloring – spontaneous decoloration in darkness'. Polymeric glasses manifesting photo-induced triplet-triplet absorption and singlet-singlet blooming are used in optical emission modulators. The photochromic materials capable of reversible photochemical transformations (spirooxazines, metal dityzonates, fulgides), are applied in production of sun protecting glasses.

The photochromic elements registering optical information have high resolution (≤ 1 nm), enable to receive the image very quickly (≤ 10^{-8} s) under the effect of light without development, store data for several years (Barachevskii 1998).

The prospects of photochromic materials are associated with the FBS development eliminating their main drawbacks: low cyclicity of phototransformations which occur in organic substances as a result of photochemical and thermochemical reactions; thermal instability of photo-induced state of majority of photochromic materials.

Holography is the technique of obtaining images based on the wave interference. Hungarian physicist D. Gabor was awarded to the Nobel Prize in 1948 for its creation. The techniques essence can be explained with the following example. The reference wave from the light source is directed to the film together with the 'signal' wave dissipated by the object. The pattern appearing during interference of these waves on the film is called hologram. When the hologram or its portion is illuminated by the reference wave, the three-dimensional image of the object appears. The FBS function of this smart system is fulfilled by the hologram property: the density distribution of photographic image in the body of light-sensitive material modulates the distribution of light oscillations in the standing wave.

Holography is used to recognize and identify images observed on the hologram, for data encoding, including those of the securities, and in devices of three-dimensional television (Denisyuk 1981).

Metamaterials implement a new paradigm of science and technology physically incarnating the folklore images of the invisible cap and invisible cloak. The metamaterials (MM) are man-made systems the size or the period p of structural units of which are much less than the wavelength λ of external emission: $p<<\lambda$. The field of this emission passes through the MM as if it were homogeneous medium. The uniqueness of the MM physical properties are the low values of dielectric and magnetic permeability and that the refraction index of MM can acquire negative values (Caloz 2009).

Fig 2.30 shows the diagram of refraction of the plane light wave passing from dielectric 1 into usual dielectric 2 and dielectric 3 with the negative refraction index (Elefthereades 2009). The flows of electromagnetic energy (Poynting vectors) S_2 and S_3 as well as the wave vector k_2 are directed from the interface between dielectrics. The unusualness of dielectric 3 is that the vector k_3 is directed towards this interface.

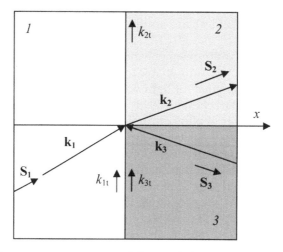

Fig 2.30. Light wave propagation in different dielectrics: *1* and *2* – usual dielectrics , *3* – dielectric with negative refraction index. S_1, S_2, S_3 – Poynting vectors, k_1, k_2, k_3 – wave vectors, k_{1t}, k_{2t}, k_{3t} – tangential components of wave vectors (Elefthereades 2009)

The metamaterials boom began after publication of the fundamental work of Russian physicist V.G. Veselago about the specific electrodynamic properties of substances with negative values of dielectric and magnetic permeability (Veselago 1967). Such materials are devoid of: the Doppler effect (the wave length change observed when the wave source travels in respect to the receiver), the Cherenkov-Vavilov emission (appears when the charged particles move in the substance with the speed exceeding that of light in this substance), the Snell law (when the light refracts at the interface between two transparent media the $\sin\alpha/\sin\beta = const$ where α and β are the angles of incidence and refraction), the effects of convergence and divergence of light beams in convex and concave lenses.

According to the theory of dissipation of electromagnetic waves, to conceal from the observer view the material object means to reduce the area of dissipated light flow to zero in the ideal. The masking with the MM is not identical with the *Stealth-technology*. The latter is directed at the reduction of radar visibility of the military objects (aircrafts, ships, defense technology). The classic example of the *Stealth* implementation by the navy is the Swedish guard ship of the 'Visby' class (Stewart 2002). Its hull is shaped with large flat sheets inclined at acute angles to conceal from view the ship typical profile. The hull is not magnetic and is made from sheets of sandwich structure: PVC inside, carbon-fiber reinforced polyester outside. Radars and infrared detectors can't locate the ship. The aviation uses for this purpose the composite coatings based on ferromagnetics simultaneously absorbing radar, infrared, laser, X-ray, acoustic and other types of emission. These coating restructure under the effect of emission, purposefully modifying the processes of particles emission and reflection. The *Stealth*-technology minimizes the emission power reflected by the object and coming up to the probing radar, that is why the object is seen mostly from the side and from the back (Alitalo and Tretyakov 2009).

So far about a dozen of techniques have been developed to conceal the object from electromagnetic and acoustic waves (Tretyakov 2003, Shalaev and Sarychev 2007, Capolino 2009). Diagram of one of them entitled 'transformation of coordinates' shown in Fig 2.31. The electromagnetic emission penetrates inside the masking 'cloak' of MM and envelops the masked object making it invisible.

For this purpose, the micro- or nanorings arranged at geometrical order are introduced into the surface layer of the masking material (Caloz 2009). Each of them consists of thin wires comparable in thickness with the wavelength of the incident emission, and divided in two parts annular resonators (Fig 2.32). The electromagnetic oscillations of light waves excite alternating electrical current in the resonators turning each nanoring into nanomagnet of which polarity changes with the frequency $\nu = 500$ GHz. Thanks to this, the MM

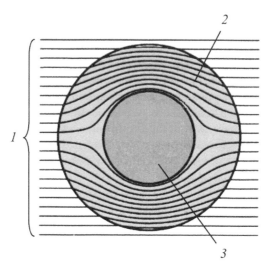

Fig 2.31. Diagram of 'transformation of coordinates' technology: *1* – external emission, *2* – masking shell, *3* – concealed object (Alitalo and Tretyakov 2009)

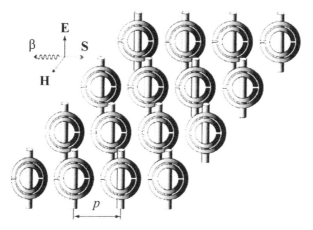

Fig 2.32. Diagram of resonating elements arrangement in the surface layer of the 'cloak' made from MM: β – electromagnetic emission, **S** – Poynting vector, **E** and **H** – vectors of electric and magnetic fields intensity (Caloz 2009)

strongly influences the magnetic and electric components of the incident light waves. The nanomagnets transfer the energy to each other creating secondary electromagnetic oscillations over the surface of the 'cloak'. The oscillations serve as conducting medium for incident light waves directing the latter over the trajectory enveloping the 'cloak' surface. If the MM eliminates all reflections and shadows, the object under this 'cloak' will become fully invisible in UV and radio bands.

A similar MM has been developed in South Korea (Bradley 2011). Metallic electromagnetic resonators are mounted into the polyimide film, each with the area 60 μm^2. Irradiation of this screen with light of THz frequency induces electrical polarization and the polyimide dielectric permeability grows. The screen refraction index augments as the clearances between resonators reduce.

Crystals of synthetic calcite (56 % CaO and 44 % CO_2) are a suitable material to fabricate the masking 'cloak' because they possess the property of light double refraction. The calcite screen conceals the objects of several mm high from red and green laser radiation and incoherent white light (Bland 2011).

Paper (Rainwater et al. 2012) reports about development of plasmonic MM (plasmon is a quantum of plasma oscillation). The MM contains thin strips of metals (Cu, Au, Ag) absorbing light and transforming it into heat and electromagnetic emission. Plasmons appear on the strip surface due to the collective oscillations of electrons emitting energy in a state of light waves. Due to this fact, the MM 'cloak' 'knocks' together the scattered waves issued from the objet which it conceals. The band of maximum dissipation of scattered emission corresponds to the frequency 2.7÷3.8 GHz. The object is invisible at any angle of observation and at any distance between the 'cloak' and the object.

So, the masking and invisibility of condensed bodies in electromagnetic and acoustic waves ceased to be science fiction. This idea is implemented in several types of MM which are smart periodic structures creating own electromagnetic field due to the energy of irradiation. It causes the incident light flow around the object to be concealed under the FBS control adjusting the flow deviation from the initial direction. The challenging problem of the materials science is to overcome the fundamental limitations on the widening of frequencies band where the objects are invisible and creation of MM which operate reliably in the band of optical emission.

(ii) **The photofunctional** materials and systems are sensitive to light. Their responses are: the photoeffect, in other words, the phenomena connected with the release of electrons from condensed bodies (photo-electronic emission, emission of gamma rays), photoconductivity of semiconductors, and photochemical phenomena (photosynthesis).

A_4B_3 are smart optical systems in which the received light signal is corrected by the FBS compensating its random distortions with the deformation mechanism.

The adaptive optical systems correct the receiving light signal by dynamically controlling the wave front shape to exclude the effect of noise influence to the limit resolution of watching instruments, to the light focusing on the receiver or target, etc. The adaptive optics began to progress in the 1950s when the problem was set to improve resolution of telescopes when atmospheric turbulence appeared to distort the received signal. The disturbance was eliminated by restoring the distorted wave front of received emission using the auxiliary light reflected by the mirror of adjustable shape. At first the set of conventional movable mirrors was used, afterwards – the flexible film mirrors which shape and curvature were easily adapted by deformation. The transition from the adaptive to the smart optical system took place after introduction of the FBS, the function was performed by computer equipped with a unit of atmospheric inhomogeneity registration. In line with these data, the computer following the assigned program corrects the shape and curvature of the wave front of auxiliary emission which eliminates the influence of atmospheric turbulence and lens aberrations. The phase-conjugated waves are used to compensate distortions of the wave front in powerful laser amplifiers. In a number of cases, the holographical methods are used to correct the wave front (Hardy 1977).

The taxons $A_4B_{13,16}$ cover the following photo-functional systems.

Instruments of pulsed photometry serve to measure the optical parameters of condensed bodies (the reflection and refraction coefficients, etc.) using the pulsing light flows with the pulses lasting less than the period of their repetition. They appeared in the 1890s with the aim to study the flashing lights (beacon, signaling) and are still in use due to the broad application of laser engineering.

Sensitive elements of non-linear crystals are used to detect the nano- and picosecond light pulses emitted by lasers. When they are affected by light flows of high intensity, some fundamental laws of classic photometry are not fulfilled in them. For instance, the coefficients of transmission and crystal spectral sensitivity become dependent on the emission intensity (Volkenshtein and Kuvaldin 1975). This property of smart crystals is the base of FBS which made it possible to receive the emissions of high density using the receivers with high time resolution and broad dynamic band. The performance of modern digital computer engineering correlates with the duration of laser light pulses that enables to process the photometric data in real-time mode.

The laser thermonuclear fusion is the reaction of coalescence of light nuclei into heavier ones initiated by laser emission. It concentrates the light energy within a small volume of thermonuclear fuel (within the area $<10^{-6}$ cm^2) within a short time interval ($<10^{-10}$ s). This provides the highest reached so far controllable energy release ($10^{19} \div 10^{20}$ W/cm^2), compression to high density (10^6 g/cm^{-3}) and fuel heating maintaining the thermonuclear reactions behavior. The laser application obviates the necessity in energy consuming system with plasma magnetic confinement because the laser emission provides the inertial confinement: the so-called ablation pressure appears at the fuel evaporation boundary equal to the sum of thermal and reactive plasma pressure. The inertial plasma confinement fulfils the FBS function proving higher effectiveness of laser thermonuclear synthesis plants in closed energy cycles with efficient energy output (Basov et al. 1975).

(iii) **Light sources** with light values are to adjusted the FBS of physical, chemical or biological nature.

Laser (light amplification by stimulated emission of radiation) is a source of optical coherent emission of high energy directionality and density. The laser principle of action consists in forced emission of photons by excited quantum systems (atoms and molecules of optical active substances) as a result of external effects ('pumping') by the electromagnetic field, electronic impact, unbalanced cooling, injection of charge carriers through *p-n* junction in semiconductors, etc. The laser consists of three elements: generator of energy processed into the light energy; active element absorbing the generator's energy and returning a part of it as coherent emission; device functioning as FBS. The first can be the electrical generator, the light source, the electronic gun, etc. The active element can be in gas, liquid or solid (dielectric crystal, glass, semiconductor, etc.) phase.

The FBS in lasers is realized with the help of optical resonator. In the simplest form, it comprises two mirrors between which the active element is placed (Fig 2.33, *a*). The emitted wave is reflected by the mirrors and returns to the active element causing induced transitions. One mirror is made half-transparent for part emission output. The annular FBS (*b*) returns the wave to the active element after consecutive reflection by the system of mirrors. It can produce the generation of two meeting waves or separate the wave of a definite direction. The distributed FBS (*c*) is implemented by the active element of which optical density varies periodically along the beam path. The reflection takes place from the boundary of sections with different optical density of substances. The reflection index from each interface is rather small but, because there are many interfaces, the value of the total coefficient turns out to be considerable (Oraevskii 1990).

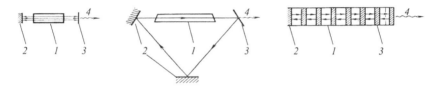

Fig 2.33. FBS circuits in lasers: *a* – elemental, *b* – annular, *c* – distributed. *1* – active element, *2* – 'blind' mirror, *3* – semitransparent mirror, *4* – generated emission

The application fields of lasers are optical communication lines, audio- and video-systems, laser technology, laser chemistry, frequency and band standards, spectroscopy, non-linear optics, medicine. As a rule, lasers are incorporated into the smart systems having the individual FBS which attaches them to the attributes of artificial intellect. Let us give three examples.

Optical location is the detection, determination of coordinates, and identification of distant objects using the light emission. The locator consists of a laser, a scanner and a photoreceiver. The laser locators assure the accuracy of determination of angular coordinates and the resolution by distance several times better than radars.

The distance R to the object is determined by measuring the delay time t of the reflected light signal arrival compared with that of the emitted signal: $R = ct/2$, where c is the speed of light. The error of R measurement is caused mostly by variations of the refractive index and atmospheric turbulence, as well as by the conditions of the laser beam reflection from the object. Nevertheless, the error of precision pulse distance meters has reached centimeters. It is due to the FBS which, taking into account the conditions of emission propagation, adjusts the duration of pulses reducing it to nanoseconds, localizes the t measurement by the pulse energy center and strobes the optical signal, that is its perception by the photoreceiver being stationary while performing discontinuous observation.

The laser optical location systems serve as the means of automatic tracking, determination of coordinates and trajectories of earth artificial satellites, as a part of the systems of spacecraft docking, make the basis of *lidars* (light identification, detection and ranging) – instruments for study the aerosols distribution in the atmosphere, the clouds form, the wind velocity (Asnis et al 1995).

The optical communication is the information transmission with electromagnetic waves of optical band. The idea of optical communication has been in use since ancient time (campfires, semaphore alphabet) until present time when emission and laser reception of broad band signals are in use. Laser data systems possess high transmission ability because of the high frequency of pulsing signals permitting to transmit a large body of information with the light speed. Low angular divergence of the laser beam assures three-dimensional security and noise immunity of information transmission.

The principle of modern optical communication is the following. The light signal carrying information is encoded into the form convenient for modulation (changes of signal parameters in time according to the assigned order), is then forthcoming into the amplifier and further into the circuit of modulator excitation. The latter sets the amplitude, intensity, frequency and phase of the output signal. The modulated laser beam is collimated (turns into a narrow beam of parallel rays) in the optical system which transmits the signal of carrier frequency to the object. There, the signal enters into the receiving optical system which focuses it at the photoreciever. The latter outputs the electrical signal by processing of which the information signal is separated.

The coverage range of optical communication (which is a usual adaptive system in the ground-based conditions) is limited by the range of direct visibility. The so-called beyond-the-horizon communication is reached with the help of FBS which function is fulfilled by the atmospheric channel of dispersion of laser signal of the carrying frequency. This system can be used in the commication lines 'Earth-space' and 'space-space' with the help of which a large body of information can be transmitted with the speed of the order 10^6 bit/s (Gauer 1989).

The fiber-optic communication lines are promising in the terrestrial conditions. The FBS function is performed there by the property of one-mode (mode – type of light wave) fiber light guides to minimize the light wave chromatic dispersion which accelerates the attenuation and broadening of carrying frequency pulses.

Laser beam technology is the material processing with laser emission. The main operations of laser beam technology are attributed to the emission heat effect. Auxiliary optical systems are used to adjust the intensity and localization of the laser beam in the assigned spot. The main advantages of the laser beam technology are: short-term action on the material, narrow area of heating, ability to treat materials in any transparent media.

Lasers are used for drilling apertures in solid, brittle, refractory and radioactive materials. They serve to produce diamond dies for wire drawing, steel and ceramic dies for chemical fibers molding, ruby clock bearings, hard alloy pieces. The drawback of laser 'drilling' is poor accuracy of apertures making due to the emission intensity fluctuations. Smart system of laser drilling should be provided by the FBS reducing the heat fluctuation of plasma particles.

Laser sheet cutting is performed by directing the laser beam and gas jet at the spot to be cut. The gas jet cools the cut edges and moves away the products of material destruction. Oxygen or air flows are used when cutting metals. Metal oxidation reduces the sheet reflection causing metal ignition, the released heat boosts the thermal effect of laser emission. The gas jet removes the melt and oxides from the cut area enhancing the oxygen inflow to the burning front. These processes induce FBS of two types mechanisms depending on the beam travel speed: i) in the mode of controllable cutting: the FBS limits the metal oxidation reactions so that the released heat is insufficient for the self-maintaining burning front; ii) in the mode of uncontrollable (autogenous) cutting: the metal melts and burns at the expense of heat of the oxidation reaction.

The laser welding can be performed within sealed vessels through transparent windows. The small heating area makes for intensive heat removal from the weld seam. This factor fulfils the FBS function due to which some metals and alloys can be welded, unamenable to welding under usual techniques, for instance, tungsten with steel or copper (Grigor'yants et al 2006).

Each of the other techniques of laser technology, such as tempering, surface microrelief producing, dynamic balancing of rotating articles, etc., uses the FBS which realize in materials the mechanisms of structural transformations, redox reactions, thermally stimulated diffusion.

Summarizing it is noteworthy that smart articles have a high level in electroengineering and electronic engineering. Namely in this domain, the first smart electronic circuits were created and the first ideas have appeared about the smart technical system as a triad 'sensor–actuator–processor'. The semicentennial development of this trend results in elaboration of materials and articles generation which surprising properties and high advantages determine the term 'smart' most exactly.

2.5. Materials and articles of medical engineering

The medical engineering is, first, the entity of instruments, devices and equipment for diagnostics, treatment and preventing diseases as well as for processing of the medical and biological information, and, secondly, the totality of articles for correcting organism state, rehabilitation of damaged and replacement of lost organs. The modern industry of medical engineering is a dynamically evolving branch of public production which incorporates the state-of-the-art of natural and technical sciences, in the first place, physics of condensed state, chemical technology and materials science, materials technology, biochemistry, biophysics and practically all medical branches.

The quality of modern medical articles has reached a very high standard. The surgery applies the grafts of virtually all human organs. Active materials are used under therapeutic treatment when the own magnetic and electrical fields of materials interact with the chemical bonds of the biological structures (Nikulin 1990). Different bandaging and sticky, film and fibrous materials are produced: antiseptic, counter-burn, analgesic, odorizing, etc.

The traditional engineering materials have mainly exhausted the potentialities of improving the medical engineering articles. Modern medicine needs biocompatible materials capable of simulating biochemical and biophysical properties of living tissues. The factors of structure, physical and chemical properties of synthetic polymers should be closed to the similar characteristics of living tissues in order to model the processes of metabolism, the mechanism of biocompatibility and adjust the biophysical organism fields using the physical fields of implants. The solution to these problems requires the development of adaptive, active and smart medical materials the application of which has social significance being targeted to strengthening health and improving quality of human life.

2.5.1. Techniques and devices of physiotherapy

Physiotherapy (PT) is the domain of clinical medicine in which natural and artificially created physical factors are used: physical fields, emission, vacuum, excessive pressure of liquids and gases, heat, cold, mineral waters, curative muds and climatic effects. These methods exclude the injury of tissues and are effective for diseases prophylaxis and medical rehabilitation of patients.

Since ancient times to the present, the main therapeutic and preventive PT principle has been and remains the method of trial and error which eliminates the very idea of application of SM and smart physiotherapeutic systems. There are objective reasons to justify the fact. The perfect intensity and duration of PT should be adjusted continuously to observe the variations in the patient organism. The exhaustive information about the state of organism can be obtained exclusively by comprehensive laboratory and clinical examination. The impossibility of achievement of this principle in real time is the main obstacle for smart PT systems developing.

The second principal difficulty consists in impossibility at the modern level of the medicinal knowledge to estimate quantitatively and transform the main organism reactions into signals determining the gravity of disease pain, inflammation and dystrophy. The pain is the psychoemotional phenomenon in which formation pain sensitive system takes part, in the higher departments of the brain and the systems of visceral (relating to internal organs) functions (Bogolyubov and Ponomarenko 1999). The inflammation is the universal adaptation reaction of organism to the effect of physical, chemical (for instance, the burn) and biological (introduction of microorganisms) factors (Sigidin 1988). The dystrophic disease originates from the disorder of nervous management of metabolism in tissues (Illarionov 2001). The quantitative assessment of these reactions in organism and its transformation into signals of physical nature is an urgent but inconceivable complicated problem solution for which is located at the interface of cybernetics, biochemistry, biophysics and medicine. Development of FBS responding to these signals by changing the intensity of therapeutic effect will become a revolution radically changing the PT abilities and significance.

The smart PT systems with the FBS provided with computing equipment which adjusts the conditions of physiotherapeutic procedures according to the state-of-art of the medical equipment. The computer transforms the signals from the sensors fixed on the patient's body (which register the temperature, the thermal expansion or contraction of the cutaneous covering, vibration of nervous and muscular fibers, the chemical composition of excretory products, the medicinal agent concentration on the skin, etc.) into the commands controlling the parameters of physiotherapeutic effect. Let us exemplify these systems.

A_1B_9 are the smart physiotherapeutic systems controlling the intensity of mechanical effects on the patient's body with the FBS which responds to the thermal expansion of the cutaneous covering.

US-therapy is the application of ultrasound (500÷3000 kHz) for the medicinal purposes. A high gradient of sound pressure (1÷15 MPa/cm) and cyclic shear deformations of the skin determine the concentration of microorganisms, activates the metabolic, immune and other processes.

Barotherapy is the application of higher or reduced air pressure for medicinal purposes. *Hypobarotherapy* is the pressure reduction over a limited skin area intensifying the flow of liquids and metabolism in the tissue resulting in vessel dilatation. Both techniques induce absorption of high doses of mechanical energy over local spots on the skin causing thermal skin expansion. The smart systems are able to adjust the intensity of mechanical effects in response to signals of tension with the sensor attached to the patient's body.

A_4B_9 are the smart PT systems with the FBS optimizing the flow of electromagnetic emission in response to the thermal expansion of the portion of the skin exposed to radiation.

Phototherapy (the light treatment) is the application of artificially induced flow of infrared, visible and UV radiation for medicinal purposes. The light energy absorbed by the skin causes hyperemia (blood inflow), more precisely – the erythema (limited hyperemia) leading to the thermal expansion of the skin on the exposed section.

Ultrahigh-frequency (UHF) therapy is the anti-inflammatory application of the electric component of high (27.12±0.16 MHz) and ultrahigh (40.68±0.02 MHz) frequency of alternating electromagnetic field (EMF) for medical purposes. Due to their long length, UHF-band waves are used to act upon large areas of a patient's body. EMF electric component is predominant in this frequencies range - it takes more than 85 % of the field energy (Bogolyubov and Ponomarenko 1999). UHF-therapy action is based on non-thermal (oscillatory) and thermal components. The latter results in stable, long-term and deep hyperemia of tissues.

High-frequency magnetic therapy is the anti-inflammatory and vasodilating application of the magnetic component of high and ultrahigh frequency EMF for medical purposes. Eddy currents (also called Foucault currents) are generated in the tissues under the action of high-frequency magnetic field. Similar to UHF-therapy, we can distinguish non-thermal and thermal components of therapeutic action. Local heating of the tissues for 2÷4 °C permeates for a depth of up to 8÷12 cm.

Super-high-frequency electrotherapy is the anti-inflammatory application of the electro-magnetic waves in decimeter, centimeter and millimeter ranges (EHF – extra-high frequency therapy) for medical purposes. Under the action of SHF radiation, heat release occurs predominantly in the surface tissues. It is accompanied by conformational change in the skin structural elements.

The afore-mentioned smart PT systems are equipped with FBS which regulates the medicinal flow of electromagnetic waves in response to the temperature sensors readings.

A_5B_3 are the smart systems in which FBS regulates the alternating current physiotherapeutic effect in dependence of the nerve and muscle fibers vibration on the area of a patient's body exposed to the action of the current.

Amplipulse therapy is the application of sinusoidal modulated currents (< 0.1 mA/cm^2) to a patient's body for medical purposes. Currents with a frequency of 5000 Hz are amplitude-modulated (amplipulse – amplitude pulsation). They generate (in the tissues) considerable conduction currents which excite the nerve and muscle fibers producing the analgesic effect.

Diadynamic therapy is the application of diadynamometric currents (half-sine wave pulses, frequency - 50 and 100 Hz, amplitude - 2 to 20 mA) to a patient's body for medical purposes. They activate metabolic processes in the tissues and relieve the pain syndrome.

Interferential therapy is the application of interferential currents for medical purposes. At summation of two currents of the same amplitude and relatively equal frequencies, their interference occurs – a stationary new distribution in time of the resultant current amplitude. Interferential currents (~50 mA) actuate rhythmic activity of vascular and visceral smooth muscles and internals. This leads to an increase in their blood supply and lymph flow.

Fluctuorization is the application of alternating currents (~1 mA/cm^2) with spontaneously changing frequency and amplitude for medical purposes. Afferent (i.e. oriented towards the nerve centers) nerve impulses which occur in the skin suppress the impulses from the pain nidus and have a local analgesic effect.

Local darsonvalization is the application of pulse alternating current (pulse duration ~100 mcs) of low intensity (~0.01 mA), medium frequency (~100 kHz) and high voltage (25÷30 kV) to a patient's body area for medical purposes. Maximum density of displacement currents and their main curative effects are achieved in the surface tissues, thus providing for treatment of neurosis, freezing injuries, trophic ulcers, etc.

Application of the aforementioned methods is accompanied by rhythmic contraction of the nerve and muscle fibers. This can be used as the basis for FBS operation which regulates the medicinal currents parameters according to readings of sensors attached to a patient's body.

Let us exemplify methods of PT which could become smart if equipped with ***means of effective side effects elimination.***

A_5B_{18} are the smart current-treatment systems whose effectiveness can be enhanced by FBS which depolarizes the tissues polarized by medicinal currents.

Electropuncture is the application of pulsed and alternating currents (25÷500 µA) to bioactive points (BAT) for medical purposes. Electrical impulses are delivered to BAT through needle electrodes, thus assuring high density of currents. Unipolar pulse currents lead to polarization of BAT-related tissues and subsequent deterioration of their functional properties. Thus, the duration of one-point single exposure to alternating currents is 15÷30 minutes, while the exposure to direct current is only 1÷3 minutes (Bogolyubov and Ponomarenko 1999). A FBS which depolarizes the tissue areas being polarized must operate in the smart electropuncture system.

A_5B_{29} are the 'smart' PT methods of current treatment, side effects of which are eliminated by FBS which initiates chemical reactions in the area of electrodes contact with a patient's body.

Galvanization is the application of direct electric current of low intensity (up to 50 mA) and low voltage (up to 80 V) for medical purposes. Indications for galvanization are: diseases of peripheral nervous system (radiculitis, plexitis, neuritis) and arthritis of traumatic, rheumatic and metabolic origin. Side effect of galvanization: electrolysis of mineral-salt-containing excretions from sweat glands. This leads to formation of acid (HCl) under the anode and alkali (KOH, NaOH) under the cathode which may cause the chemical burn of the skin.

The smart system electrodes may be hollow with microchannels which come out of the cavity to the working surface of the electrode. When the electrolysis products have been accumulated under the electrodes, FBS operates responding to pH sensors signals. As the result, solutions for electrolysis products chemical neutralization are delivered (through microchannels) to the area of the electrodes' contact with the body.

A_7B_{10} are the smart systems intensifying electrodiffusion by means of FBS which increases the diffusion flow by increasing the medium permeability.

Medicinal electrophoresis (galvanic ionic medication) is the application (to the organism's tissues) of direct current and medicinal agents (MA) introduced through the skin and mucous membranes for medical purposes. The amount of MA penetrating the tissues during the electrophoresis makes 5-10% of the total amount of the substance delivered under the electrode (Bogolyubov and Ponomarenko 1999). Increase of water solutions MA concentration above 5 % leads to onset of electrophoretic and relaxation forces which decelerate MA ions movement in the tissues (Debye-Huckel effect). The FBS is required to offset the braking forces.

State-of-the-art devices for PT where multifunctional active materials are used, meet modern standards for medical equipment. They are exemplified by the needle applicators designed to relieve pain in the muscles, joints, spinal column and normalize cardiovascular, respiratory and nervous systems.

The applicator invented by Russian doctor I.I. Kuznetsov is a flexible mat with metal needles fixed on it (Kuznetsov 1980). Ukrainian doctor N.G. Lyapko has designed *an applicator* in the form of a rubber roller with needles fixed on the surface. Reflex and mechanical impact of these devices lies in the BAT multiple acupuncture and micro-massage of skin and subjacent tissues. N.G. Lyapko believes that the stress caused by the applicator discloses the organism's potential of self-healing (Lyapko 2007).

The applicator needles may be made of various *metals, microdoses of which are necessary for human body* (Zn, Cu, Fe, Ni, Ag, Au, etc.). Intensity of the acupuncture galvanic-electrical action is regulated by the body itself. According to N.G. Lyapko, the human body is a gigantic biocolloid system consisting of electrolytes; the galvanization effect depends on the extent of tissues (skin layers, hypodermic tissue, underlying structures) saturation with the electrolytes. The galvanization effect is composed of the following: diffusion of micro-elements coming from the needles and their depositing in the human body; pH normalization of the tissue fluid; activation of metabolic processes in the tissues; stimulation of metabolic processes (Lyapko 2007). This imparts to the treatment procedure the features of a smart system, whose effectiveness is regulated by natural FBS inherent to the patient's body.

The needles may be made *of magnetic or electret material*. The applicator in the form of a heteropole magnetized resin-bonded magnetic mat with ferromagnetic metal needles (Fe, Co, Ni) fixed on its magnetic poles is used for magnetic therapy of reflex areas (Pinchuk et al. 1995). Ferromagnetic and electret needles of polarized dielectric are installed in the magnetic base (Makarevich et al. 2003). This applicator is the source of magnetic and electric fields which exert synergistic action upon reflex areas of the human body. A magneto-therapeutic mat is proposed for prophylaxis and medical treatment of osteochondrosis and radiculitis among professional drivers (Pinchuk et al 1992). The mat acts upon spine reflex areas with its magnetic massage rollers rotating on flexible axles. They have been magnetized to optimal (depending on sensitivity of reflex areas) values of magnetic induction – from $20 \div 35$ mT to $70 \div 85$ mT. Applicators of this class are multifunctional active PT systems, but they are not the smart ones. They generate constant physical fields which exclude the possibility to regulate their therapeutic action and the necessity of using FBS.

Modern medical engineering offers an extensive selection of original sensors (Reece 2007) which are simply impossible to be fully represented herein and which provide for creation of the smart PT systems. Nevertheless, that step has not been taken yet. Naturally, the ways of implementing smart-system-aided PT methods are subject of patenting.

2.5.2. *Materials for implants*

Implants are the technical products being implanted into an organism to compensate functional defects of the damaged or missing organs. Modern achievements of the implant surgery are the result of not only advanced surgical skills, but also of the materials science progress in the study of biochemical interaction between living tissues and implants. As a result, the implants came to be made of materials which (on their own, as well as products of their interaction with living tissues) do not cause the rejection reaction.

The main requirement to the implant materials is their biocompatibility. The definition of this property in the International standard (Williams 1987) appears quite vague: Biocompatibility is the ability of an implant to perform with an appropriate host response in a specific application. In other words, firstly, biocompatibility is not an intrinsic property of any material, but the characteristic of 'implant-organism' system corresponding to their condition-specific contact. Secondly, on very rare occasions an organism perceives the implant as a 'native' body. Biocompatible implant elicits the organism reaction which makes it possible to solve the set task.

The living tissue response to the implant develops as follows. Blood serum comes into contact with it first. Its protein macromolecules are adsorbed on the implant and enter into donator-acceptor interaction with atomic particles on its surface generating complex chemical compounds. The receptors of the cell membranes interact selectively with these bound particles – ligands. When the amount of the "receptor-ligand" complexes is sufficient, they send a signal to the cell nuclei which gives rise to a cascade of intercellular chemical reactions. The latter regulate the structure and functions of cells in contact with the implant (Dee et al 2000). Depending on the nature of occurring biochemical processes the following types of the tissue response to the implant are distinguished (Karlov and Shakhov 2001):

- toxic – toxic damage of the tissue, i.e. necrosis, destructive inflammation, dystrophy and atrophy, degeneration;
- bioinert – the tissue forms a fine inadherent fibrous capsule round the implant;
- bioactive – the tissue is biologically linked with the implant at the interface;
- implant dissolution – the tissue absorbs and substitutes the implant.

Most frequently, implants provoke the bioinert reaction of tissues. Connective tissue cells emerge around the implant, thus forming a boundary. From this boundary the living tissue aggressively acts against a "stranger" trying to remove it from the organism. Chemically active liquids containing enzymes (biological catalysts for substances transformation in the organism) exude from that boundary. They initiate biocorrosion of metals and even dissolution of ceramics (Nicholson 1998). Filamentary connective tissue forms a fibrous capsule which isolates the implant from the living tissue and is in state of dynamic balance with it.

The implant materials range includes practically all types of constructional materials (Karlov and Shakhov 2001, Nicholson 1998, Savich et al 2003, Petty 1991, Pinchuk et al 2006). Table 2.1 compares the strength factors of the main types of implant manufacturing materials (Petty 1991). Ceramics, metals and carbon have the highest strength which has always been a valuable parameter. However, in the 21st century, the long-continued domination of metals among materials for implants ceased to be absolute. Composites allow to simulate the bone tissue deformation-strength characteristics and, to a greater extent than other materials, correspond to biological environment. The modern tendency of the medical engineering materials science lies in the fact that "migration" of the best engineering materials into medicine is coming to an end while the process of targeted development of medicinal-purpose materials is being started (Pinchuk et al 2006). Nevertheless, it is very rare to create an analog of the living tissue whose properties depend upon distribution of biophysical potentials and filling with biological liquid.

Table 2.1 Deformation-strength characteristics of graft materials

Material	Properties			
	Yield strength, MPa	Breaking strength at tension, MPa	Relative elongation at rupture, %	Elasticity modulus, GPa
Ceramics				
Aluminum Zirconia:		550		380
- made by hot isostatic pressing		1200		200
- sintered		900		200
Metal alloys				
Protasul-10, forged	1000	1200	10	200
Stainless steel				
-after mechanical processing	750	1000	9	200
-annealed	170	400	45	200
Alloy Ti (6 %) – Al (4 %) – V	890	1000	12	105
Carbon and composites				
Gas-phase-deposited carbon		350-700	2-5	14-21
Carbon fibers				
-low-modulus		1720	0.75	380
-high-modulus		2760	1.4	240
Carbon-carbon composite with:				
-parallel fibers		1200		140
-mutually perpendicular fibers		500		60
Polysulfone–carbon composite		2130	1.4	134
Biomaterials				
Hydroxyapatite		100	0.001	114-130
Bone		80-150	1.5	18-20
Collagen		50		1.2
Polymers				
Polymethyl methacrylate (bone cement)		75	3.5	2.8
Ultrahigh molecular polyethylene (UHMPE)	23	40	500	0.5
Silicon rubber		7-10	600	0.0003

Creation of the biological-tissue-simulating smart systems is a noble task for the implant developers. In view of high responsibility related to implants implantation, they are manufactured from materials characterized by increased structural and properties stability which, generally speaking, contradicts the smart materials concept. Nevertheless, smart materials for implants have been created, and their development is underway which is proved by the examples below.

A_1B_2 are SM and smart implant systems which (in response to mechanical deformation) activate FBS initiating the implant structure change.

Smart increase of the cross-section of joint implant (JI) stems fixed by press-fitting of the bones into the marrowy canal will allow eliminate the JI loosening syndrome. For the first time that new pathological state of patients was noted in 1966 by one of the pioneers of endoprosthesis replacement Swiss orthopedic surgeon M. Muller. To solve this problem, SM with adjustable Poisson's ratio – both positive and negative – are required. Specimens of materials with negative Poisson's ratio – auxetics – expand in the plane normal to the direction of stretching and compress along the normal to the direction of compression (Konek et al 2004). The FBS in such SM is quite complex: it should respond not only to loosening of the stem, but also take into account the bone strength at biaxial stretching.

A_1B_{32} are the smart systems in which biomaterial mechanical deformation initiates the biosynthesis regulated by FBS taking into account the deformation value and direction.

Ilizarov apparatus is designed for long-term fixation of fractures, compression or stretching (distraction) of the bone tissue formed at fragments splicing (perosseous or compression-distraction osteosynthesis). It is used for medical treatment of fractures and deformations of bones. The apparatus was invented and put into practice by an outstanding Russian arthrologist G.A. Ilizarov (Ilizarov 1952). 'Ilizarov effect' is the occurrence of the newly-formed bone tissue lengthening in the field of tensile stresses. Groundwork has been laid for developing a smart Ilizarov apparatus where a computer takes control of the bone deformation (Savich et al. 2003). The FBS compares the value of emerging stresses with the deformation-strength parameters of the newly-formed tissue and expedites the healing by creating compression and distraction micro-displacements of the bone fragments.

A_2B_{32} are the smart friction units equipped with FBS initiating the friction material biosynthesis.

A human joint which operates under considerable loads for dozens years with a very low friction coefficient ($0.02 \div 0.05$) and compensation of cartilage wear represents a 'smart' friction unit. The FBS function in it is performed by the sympathetic nerve center which regulates permeability of blood capillaries, production of lubricating synovial liquid (SL) in the joint, its circulation in the area of cartilage friction, etc. [55]. *Joint implants* (implanted prosthesis of movable bones joint) differ from the natural joints, mainly, by lack of cartilages – anti-friction microporous bodies filled with SL. When a healthy joint is loaded, SL is exuded from the micropores on the friction surface areas which are under the heaviest load and provides for their lubrication. Creation of an artificial cartilage is the first step in bringing endoprostheses closer to the synovial joint smart system. It represents a microporous layer (Fig 2.34) formed on the joint implant parts made of ultrahigh molecular polyethylene (UHMPE) which has proved its reliability in the endoprostheses friction pairs since 1960s. The micropores of artificial cartilage may be filled with medicinal substances expediting the healing of operative wound (Pinchuk et al. 2006).

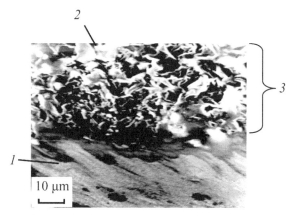

Fig 2.34. Cross-section of UHMPE artificial cartilage: *1* – UHMPE initial structure, *2* – friction surface, *3* – microporous layer

Joint surgery is not just removal of pathologically changed tissues, but also a procedure which violates the natural distribution of the joint's biophysical field. The biofield defects can be compensated by electric polarization of the artificial cartilage. The polarization charge which has been formed in the dielectric part of joint implant remains in the biological liquid for a long time during friction. It increases adsorption of SL and its tribodestruction products on the artificial cartilage surface (Pinchuk et al 2001b). The adsorption layer provides for boundary lubrication and prevents the polymer part fatigue wear which develops by a mechanism of subsurface cracks formation.

Unipolar joint implants work in friction pairs with the natural cartilage. The implant spherical head traumatizes and gradually destroys the cartilage. To extend the reconstructed joint 'lifetime', an artificial UHMPE cartilage is formed on the head; the cartilage micropores are filled with medicinal substances (chondroprotectors) expediting the cartilage physiological regeneration (Pinchuk et al 2011). The chondroprotector is exuded from micropores exactly in those areas of the natural cartilage which are in contact with the head and are deformed by it. This is a signal to actuate the natural FBS of the joint which initiates the cartilage tissue regeneration. Currently the smart implants with the natural cartilage undergo clinical trials in Belarus.

A_1B_3 are the implants in which FBS-regulated stresses occur under temperature change.

Implants with memory effect are used in many clinics around the world. Russian and US experts have created original titanium nickelide dental implants with the elasticity modulus $E = 50 \div 75$ GPa which is close to E value of the bone tissue (Gyunter et al 1998). However, the unusual properties of the implants with memory effect do not constitute grounds for their attribution to SM category. It is practically impossible to regulate the temperature (as an instrument to activate the memory effect) of a human organ which the implant is implanted into. There is a need for materials implementing the memory effect under another energy deposition, e.g., under the action of radiation. Implants made of these materials could change their form under FBS control with a computer managing the radiation intensity according to the signal of sensors which register the sick organ condition.

A_8B_{12} are SM formed as the result of chemical reactions which thermal effect is controlled with FBS optimizing the material structure.

Bone cement is used to stick implants to the bone tissue. It is produced by mixture polymerization of powdered methyl methacrylate and liquid polymerization initiator. A unique structure of the cement in the form of two spheres made of calcium phosphate flake-like crystals (Fig 2.35), demonstrates high mechanical stability.

The polymerization reaction is accompanied by the release of large amounts of heat: at curing of the bone cement standard (for the hip joint implant stem or socket) dose of 4 cm^3, its temperature reaches 100 °C in ~15 minutes (Savich et al 2003). Such temperature is fatal for the living tissues. Evidently, there is a need to develop smart cement the preparation of which the competitive chemical reaction would occur: the first one (exothermic) with the heat release and the second one (endothermic) with the heat absorption. The product of the second reaction shall complement the properties of the polymethyl-methacrylate-based basic bone cement.

A_9B_2 are SM which properties are directional changed as the result of biological environment influence by the control of FBS which regulates the material structure change.

Bioactive ceramics coatings – tricalcium phosphate $Ca_3(PO_4)_2$ and и hydroxyapatite $Ca_5(PO_4)_3OH$ – are designed for the implants osteointegration (Pinchuk et al 2006). Hydroxyapatite (HOA) chemical and crystal

10 µm

Fig 2.35. Typical microstructure of 'CPC' bone cement after curing

structure is identical to the bone tissue structure (Geesink et al 1988) which provides for formation of strong bonds between the bone and the coating. Perfection of technologies of HOA plasma spraying onto metal implants resulted in the growth of the coating adhesion strength from 5÷7 MPa to 12÷14 MPa (Savich et al 2003). Owing to this fact, good early operation results had been obtained. However, 7÷9 years after HOA-coated implants installation, first indications of their instability and loosening are revealed. Practically complete HOA dissolution in the bone tissue results in formation of a gap 0.5÷1.0-mm wide between the hip joint implant stem and the cannon bone (Semlitsch 1992). In view of this, it is vital to create 'graft-coating-bone' smart system where the bone tissue does not only 'eat away' the coating, but also comes into contact with the implant without reducing the connection rigidity. To solve the task, it is necessary to implement a FBS regulating HOA resorption (resolution) along the coating area.

Resolving sutural materials disintegrate in the organism under action of biological liquids and enzymes within two-three months. The suture must be strong in the period of the surgical wound healing and decompose after coalescence of tissues which it unites (Shishatskaya et al 2002). The dissolution rate of sutures varies depending on their nature (protein-based or synthetic polymer-based) and aggressiveness of the patient's organism biological environment. In the 'smart' suture materials this factor must be regulated by FBS depending on the degree of the suture-connected tissues coalescence. Currently, this task is deemed neither vital nor having reliable ways of solving it; however, its importance for surgery is increasing over time.

Similar problems are typical for *dissolving substitutes of bone transplants.* Works on them were started in the second half of the last century and are underway until now. New strategies: nanomaterials self-assembly; initiating bones formation by gene engineering methods; use of platelet-rich blood extracts and non-physiological doses of hormonal MA, etc. – are still not effective enough (Bohner 2010). Creation SM of this class is the task for the immediate future.

A_9B_{17} are the smart electret implants equipped with FBS which regulates the motion intensity (in the electret field) of the charged particles which displacement promotes the medical treatment.

Electrical stimulation of osteoreparation, i.e. growth and rehabilitation of bones, became it is possible in orthopedic surgery after detection of biophysical potentials and electrokinetic phenomena in the bone tissue (Karlov and Shakhov 2001). This method is used for treatment of bone fractures, osteoporosis and shortening of limbs. The field of Ta_2O_5 or polytetrafluoroethylene electret coatings on tantalum holder of bone fragments speeds up (by $1.5 \div 2.0$ times) the rehabilitation and functional recovery time of the limbs in case of closed, open and gunshot fractures (Morgunov et al 1994). The mechanism of the electret field effect on the osteoreparation process can be viewed as follows (Black 1994). Within the conducting but screening medium of the organism, the electret field spreads to a slight distance and therefore affects only the neighboring implanted cells. More distant cells are activated by the nervous conductors, via a forwarding of the excitement. This activation may spread to considerable distances and stimulate osteoreparation.

It would be logical to control the osteoreparation process by directional change of the electret field, however it is constant and cannot be adjusted. The FBS function in the biological environment of electret implant is performed by the natural ability of the organism to optimize electrical conductivity of tissues, nerve fibers, blood and lymph capillaries.

A_9B_{18} are the smart implants whose biocompatibility is controlled by FBS changing the electrostatic potential on the implant surface.

The effect of electrical fields on biocompatibility of the implants is one of intensively explored directions of medicinal techniques today (Feast et al 1993). Biocompatibility is determined by the tissue liquids adsorption on the implant, i.e. by correspondence between the implant surface potential and bioelectrical potential of the surrounding cells. The so-called 'polymer brushes' represent a monolayer of macromolecules adsorbed on a solid body surface. They are oriented normally to the surface, densely packed and have one end of their chain adsorptive linked with the solid body. The surface of implant with the 'polymer brush' acquires a charge which is similar in sign to that of the 'free' end of chains (Nagasaki and Kataoka 1996). The 'brush' electric adsorption potential defines, to a great degree, the bonding of biological environment components (proteins, aminoacids, cells) with the implant. Thus, the surface of any implant covered with the 'brush' of polyethylene glycol molecules is hostile towards protein adsorption but is highly compatible with blood (Hurris 1992).

The FBS function of regulating biocompatibility of the implant equipped with the 'polymer brush' is performed by the natural ability of the patient's organism to directionally change pH of biological liquids (from 2 to 10) and temperature (from 30 to 40 °C) of an organ which the implant is being implanted into. Density of macromolecules in the 'brush' and, consequently, the implant biocompatibility depend upon these parameters (Nagasaki and Kataoka 1996).

Robot-assisted surgery instruments were created at the end of the 20th century because the doctors were striving to perform surgical procedures through natural openings in the human body or by doing minimal (~ 1 mm) incisions in the tissues. Miniature surgical instruments usually include a source of light, receptacle for excised material, cutting, gripping and suturing units (Pasic and Levine 2007]. The procedure performed with their aid allows to reduce the shock experienced by a patient waiting for surgery, decrease the risk of surgical traumatism, loss of blood and the organ postoperative rehabilitation period. The robot control system includes electronically controlled electromagnetic or pneumatic drive. This allows the surgeon to perform the operation with higher degree of freedom and accuracy (Fernandes and Gracias 2009).

Information on these smart systems is given at the end of the paragraph, since their attribution to a specific taxon depends on the nature of the robot controlling signal. The robot operates according to the computer-assigned program which is corrected FBS according to signals of biochemical, physical, mechanical or other nature which come from the organ under surgery.

2.5.3. Target-oriented drug delivery

Medicinal agents (MA) entering the body go a long way before they reach the effector organ or tissue. At oral intake, only a small amount of MA reaches the target. Many drugs start to disintegrate already in the gastrointestinal tract; the MA amount absorbed into the bloodstream goes through the liver where it is subjected to biochemical transformation and retained in the tissues, etc. The amount of MA dose absorbed by the effector organ determines the degree of MA bioavailability. The injected MA reach the bloodstream quicker and have a higher degree of bioavailability. At subcutaneous administration, the rate of MA delivery to the diseased tissue is lower, however, due to lack of disintegration in the place of introduction the bioavailability is as high as at MA injection (Varfolomeev and Gurevich 1999).

The target-oriented delivery presupposes the MA placement directly into the effector organ by-passing other systems and tissues of the organism. This medical technology has been developing intensively since the end of the 20th century and is currently the main trend in pharmacokinetics whose purpose is to achieve MA dose optimization to increase therapeutic and reduce toxic effects. In principle, the problem statement itself presupposes the use of SM and smart medical systems.

A_1B_{32} are smart systems for mechanical delivery (into the human body) of MA which propagate in the disease-affected organ by concentration diffusion mechanism.

Subcutaneous administration of MA microamounts is exercised with hollow needles (with diameter up to 10 mcm) made of metals, ceramics, glass or polymers. Microneedles are assembled into units each containing several hundreds of them. This method is an alternative to the technology of medicamental therapy with acupuncture employed for about 100 years. Microneedles are entered to a depth of $10 \div 15$ mm from corneal layer which causes no pain as there are no nerve endings there (Cordelia 2004).

The systems of subcutaneous administration of MA microamounts are typical active systems. They become smart ones when equipped with the FBS regulating MA diffusion kinetics. Such FBS functions can be performed by microneedles of biodegradable polymers whose decomposition rate in the organism optimizes MA release process.

A_6B_{22}, A_6B_{25}, A_6B_{32} are smart magnetic nanoparticles whose movement in the organism under external magnetic field gradient action is controlled with the FBS; the latter initiates their fixation on the disease-affected cells by means of adsorption and immune reactions.

Magnetic nanoparticles – the diseased tissues markers – are made of Fe, its oxides, Ni or Co. They are isolated with coatings of Au or bioinert polymers; functional chemical (proteins) or biological (antibodies) agents are grafted to the coating (Fig. 2.36). The coating prevents aggregation of particles, while the grafted groups (antibodies, oligonucleotides, peptides) serve as the FBS elements.

Sufficiently small particles ($500 \div 100$ nm) are injected into the blood stream. They do not obstruct micron-sized capillaries, penetrate through their walls and are adsorbed on cells affected by cancer under the action of the FBS which initiates intermolecular adhesion of cells and grafted groups. The particles form (on the affected tissues) clusters which are registered by the magnetic resonance method (Gould 2004). This provides for detection of individual cancer cells in the early stages of the disease and not at the stage of a large tumor emergence.

The magnetic nanoparticles can carry MA molecules, so the particles acquire a special function of the tissues therapeutic processing. Chemotherapy is the lamentably well-known unpleasant side of the cancer treatment with highly-toxic agents hazardous for healthy cells. Smart nanoparticles are retained only on the cancer target owing to adhesion interaction between it and the grafted proteins and antibodies; they focus

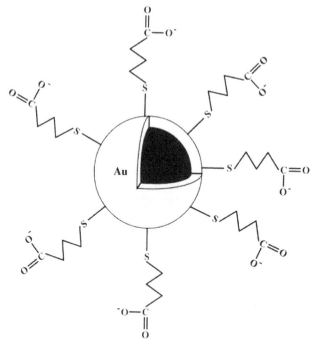

Fig 2.36. Structure of magnetic nanoparticles with functional groups attached through Au-S bonds.

the magnetic field on the target by literally 'pushing' the MA molecules towards the malignant cells (Gould 2004).

A_8B_{13} are smart carriers of chemotherapeutic agents which are adsorbed (using the grafted groups) on the malignant target and release the MA under the action of external radiation.

Carbon nanoparticles have long been used as a means of MA delivery to the cells affected by cancer. Discovery of fullerene (in 1985), carbon nanotubes (in 1991) and graphene (in 2004) has practically immediately determined their application as a transport means in chemotherapy. It is believed (Liu et al 2011) that they interact with malignant tissues by mechanism of endocytosis, i.e. they are absorbed by the cells. The MA are preliminary introduced into the free volume of fullerenes and nanotubes or grafted to the graphene lattice with delocalized p-electrons (Fig 2.37).

The attractive property of carbon nanoparticles is their optical activity in near-infrared region ($\lambda = 700 \div 1000$ nm). External optical signal releases MA from the particles absorbed by the cells and actuates the photothermal tumor ablation (Liu et al 2011). In view of this, a question of long-term nanotubes biocompatibility with the human organism has arisen.

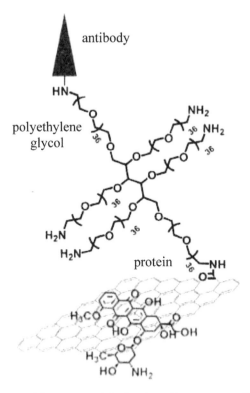

Fig 2.37. Graphene oxide nanoparticle for detection cancer cells and MA target-oriented delivery to them

Pre-clinical experiments on animals (from rodents to primates) in many countries revealed weak interaction between nanotubes and immune, reproductive and nerve systems. The result of these studies can be summarized as follows: toxic effect on the organism of nanotubes and tattoo inks employed for thousands years has been identified and proves that nanotubes are practically harmless (Hara et al 2011).

A_6B_{23} are the smart MA-carrying nanoparticles which are used to introduce MA into the human body and fix them in the required location with the help of external magnetic field; then MA are released from these locations under the activating action of concentration diffusion.

Wireless-controlled nanorobots on the base of carbon fibers (length ~500 nm; diameter ~85 nm) are the modern means of target-oriented delivery of drugs to malignant cells (Zeeshan et al 2011). They are administered intravenously, their movement is controlled and they are fixed in the affected tissue with the help of external magnetic field. There nanorobots are activated, MA are released and diffused into the cancer cells.

2.5.5. Tissue growth

Transplantation of organs and tissues as a surgical method has been known since antiquity to modern times (Shvarts et al 1999). Tissues belonging predominantly to a patient are transplanted. This procedure dictates the necessity to perform another (additional and quite painful) surgical operation on tissues extraction. Alternative solutions are transplantation of tissues taken from the tissues bank (banks of bone tissues exist since 1970s, banks of soft tissues are not available yet) and growing of tissue implants by living cells multiplication. The latter solution, which is called the tissues engineering, presupposes availability of a special scaffolds where a tissue embryo could develop by assuring biocompatibility and sufficiently long 'lifetime' of the tissue implant.

Creation of such scaffolds (the process was started in the 1990s by several enthusiastic researchers) has grown in the 21st century into an important direction of the materials science development; today it is a dominant subject in books and magazines on biomaterials (Ma and Elisseeff 2005). The scaffolds form the histoarchitecture which determines biomechanical and biochemical characteristics of the growing tissue surface layer. Biodegradable scaffolds are transplanted together with the grown tissues, and they exert influence upon biological coherence of a new tissue with the multitude types of the surrounding cells (Perez-Castillejos 2011).

The main direction of the tissues engineering is the genetic engineering – a division of molecular genetics dealing with purposeful obtaining *in vitro* of new combinations of genetic material, which is capable of multiplying in the host cell and synthesizing the final metabolic product. One of the directions of obtaining genetically new tissues is called cloning – vegetative (asexual) multiplication of cells derived from one stem cell (clone). The clone is obtained using stem cells which are found in human ovum at early stages following fertilization. DNA spiral is extracted from the stem cell and replaced by DNA from the patient's cell. When cultivating the clone *in vitro*, human tissues can grow genetically compatible to the tissues of a person who 'gave' his DNA. This imparts the features of artificial intellect to the tissues engineering products.

These systems correspond to A_9B_{32} taxon: environment (*in vitro* it is the cultural medium and scaffold which the cells have been inseminated onto, *in vivo* it is the patient's organism) exerts biological action upon the tissue embryo by initiating its growth. The FBS function is performed by the natural ability of the living cells to biosynthesis under certain conditions.

Scaffolds on the base of natural tissues are used in many medical technologies. Examples of their application in orthopedic surgery are cited below.

Bone grafting is a set of surgical procedures for locomotor apparatus rehabilitation using plastic materials in the form of bone tissue. About two million patients with diseases and traumas of musculoskeletal system are annually treated with this procedure (Bohner 2010). The bone tissue belonging to the patient (autotransplant) is used quite rarely, while living tissues belonging to other people (allotransplant) or tissues of some other origin (xenotransplants) are used more frequently. On all occasions the implanted tissues serve as a matrix where amplification of human cells takes place.

Allo- and xenotransplants can initiate immune responses facilitating the disease transmission. As a precautionary measure against that, the transplants are subjected to special treatment: chemical, low-temperature and X-ray irradiation (Pinchuk et al 2006). Deproteinization, i.e. extraction of proteins from the bone tissue, reduces the risk of the 'alien' macromolecules appearance in the patient's organism. Bovine bone-based xenotransplants – deproteinized 'BioOss' and heat-treated (under $T \approx 1000$ °C) 'Endobon' – are used in clinical practice (Bohner 2010).

The transplant engraftment starts with formation of blood vessels in the area of contact with the host bone. The process is controlled by the organism immunes responses, thus, it is limited by the transplant layer of not more than 1÷3 mm. After the blood-fed osteogenic cells have formed the living bone tissue on the transplant matrix, it transforms into a biocomposite consisting of viable and necrotic bone tissues (Scarborough and Griend 1991).

There is no doubt that such systems are smart ones. The FBS function in them is performed by the natural ability of the bone tissue to regenerate which is limited by the patient's age, past medical history, distribution of blood vessels in the transplant, etc.

Articular cartilage rehabilitation is exercised by transplanting the mature cells of the patient's cartilage tissue (chondrocytes) into the cartilage defects. This procedure is standardized at international level. During knee joint arthroscopy, a bioptate (biopsy slice) of the cartilage tissue is taken from the non-load area of medial condyle. Then the bioptate is crushed, chondrocytes are filtered, mixed with the patient's blood serum and grown in the nutrient medium. 2÷3 weeks later, a suspension of chondrocytes is formed which is used to close the cartilage defects (Brittberg et al 1994).

The newly formed cartilage is formed on the subjacent bone which serves as a scaffold for growing autochondrocytes. Nevertheless, it differs slightly in collagen structure from the initial cartilage (Mandelbaum et al 1998). Consequently, the scaffold does not play the core role in the cartilage tissue growth. Even in the own organism internal environment, the growing cells are influenced by the factors (presumably, immune ones) stronger than the scaffold biophysical influence.

It is difficult to imagine an FBS which would allow to regulate the natural processes of implant survival and transplant tissues growth in biological environment of the living organism. The process can be influenced somewhat by medical technologies which should be related to external actions but not to FBS.

Artificial scaffolds have specific advantages which can be attributed to the following. Due to a vast variety of the tissue transplants and factors influencing their implantation into the patient's organism, it is very difficult to define the features of an ideal tissue-growing scaffold. Allotransplants have set high standards for scaffolds: increased activity on biological ligands bonding, 'actuation' of growth factors, sensitivity to tissue proteases (enzymes) acting as a catalyst for breaking the peptide bonds in intracellular proteins (Patterson et al 2011). The artificial scaffolds have other advantages: i) relative simplicity of their formation and methods of physical and chemical characteristics control; ii) possibility of the biological material inoculation under mild physical conditions with negligible loss in its viability; iii) elimination of the problem of the scaffold genetic cleaning and threat of pathogenic agents transfer into the patient's organism. Examples of artificial scaffolds which has been successfully applied in the tissues engineering are cited below

Polymer hydrogels – cross-linked polymers on the base of hydrophilic macromolecules are able to have a balanced and reversible expansion in water solutions. Biocompatible hydrogels simulate the soft tissues viscous elasticity, are soluble in biological liquids and form covalent bonds with the living tissues; the mass transfer in them occurs by the mechanisms of diffusion and water phase microdisplacement.

The hydrogel scaffolds for the tissues engineering are made of hydrophilic synthetic polymers and polysaccharides – polyethylene glycol, polyvinyl alcohol, polyhydroethyl methacrylate, fibrin, collagen, etc. The scaffolds have fibrillar structure in the form of a fibrous grid creating the possibility for cells migration of the growing tissue (Fig 2.38).

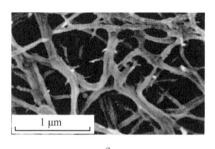
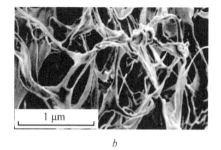

a   b

Fig 2.38. SEM images of the hydrogel structures on the base of: *a* – fibrin, *b* – hyaluronic acid

The artificial intellect features are demonstrated by the hydrogel scaffolds engrafted (for enhancement of biological activity) with functional molecules – fragments of peptide and protein macromolecules. They interact with the cells by regulating their migration in the scaffold. Owing to dipolar moment availability in the hydrogel fibrils, it is possible to control the migration with a strong magnetic field (Patterson et al 2011). The factors of growth performing the FBS function (i.e. substances which promote the cells division) are introduced into the smart scaffolds of collagen hydrogel.

Scaffolds of polyoxymethylene nanofibres are formed by melt-spinning. The polymer solution is squeezed through a spinning jet. At extrusion under own weight, the fibers are electrically polarized and subjected to an air flow. It removes the evaporating solvent and tangles the charged fibers deposited in the earthed scaffold. The excess charge is relaxed while the residual charge (the fibers electret charge) – exerts influence upon the crystal structure formation and macromolecules orientation in the fibers (Agbenyega 2011). It performs the FBS function in the process of migration of the tissue-transplant-forming cells.

Bioceramic coatings which are applied onto the implants to stimulate the bone tissue implantation (see 2.5.2, A_9B_2 taxon) are, as such, a scaffold which activates development of the bone tissue cells (osteocytes). The growing tissue surface layer is being formed in contact with them. The coatings of calcium phosphate (CaP) are biocompatible. They stimulate osteocytes growth and have microporous structure which is convenient for the FBS implementation in such biosystem. Its function is performed by therapeutic agents, proteins and growth factors introduced into the micropores. They control the cells structure of the surface layer, suppression of infections, the graft's 'lifetime' (Campbell 2003).

In conclusion, we would like to draw the attention to the ethical aspect of implementation of the genetic engineering smart systems. Uncontrollable distribution of the 'engineered' organisms may have serious consequences for life on Earth. The church blames these activities as ungodly. Nevertheless, in 1982 the European Parliament on its meeting in Strasbourg adopted recommendations on legitimacy of the genetic engineering techniques application to human beings. In 2001 the British Parliament adopted a law which permits cloning cells from a human to obtain transplants; the law regulates the donor rights and their mandatory consent for conducting a study. Similar laws have been adopted in a number of other countries However, the social aspect of this problem remains controversial, and genetic engineers perform their duties under heavy load of public opinion (which is not always favorable) and increased personal responsibility for the final result.

The use of medical equipment shall comply with the slogan 'First do no harm!', thus the materials science of medical engineering is one of the conservative trends in the science of materials. Nevertheless, unique SM and smart systems where the FBS functions are performed by the living cells and tissues have been created in this field of science. Perspectives of the smart medical equipment depend largely upon the successes in genetic engineering the projects of which receive mixed judgments by the public.

2.6.   Means of packaging

Packaging is an arrangement, most often a container, which raw materials, their processing wastes or finished products are placed. The packaging serves three basic functions: i) protection of materials and products against damage and loss, ii) protection of environment against pollution with the package content, iii) product maintenance during movement, i.e. during transportation, warehousing, storage and sales (State standard 1986). The modern packaging is the factor for increasing the product competitiveness, thus the number of its functions has increased dramatically leading to enhancement of the packaged goods safety and reliability.

The variety of interrelated functions of the modern packaging has determined its trend to become a smart system. The first problem which had to be solved on this way was the rapid increase in the volume of mandatory information on the product. Now it is necessary to provide data on allergenicity, freshness, health benefits, ethics of the foodstuff production, evidence of the fact that the packaged product is not a counterfeit. This had initiated application of new forms of information on the package: sound signals, flavor, color change as an integral indicator of the product deterioration over time and temperature. In sophisticated packaging forms, physical and chemical sensitivity of its material is complemented with electronic data processing and the result presentation on disposable displays remotely informing the users on non-stable condition of the product (Butter 2006). It was facilitated by the prevailing use of polymer materials in the packaging industry.

The diagram on Fig 2.39 shows the application fields-wise distribution of plastics on the example of economically developed European country Austria. The data are as of the end of the 20th century, however over the past years there has only been a slight change towards the increase of the share of packaging-purpose plastics. The examples cited below show that just the plastics constitute a technological basis for the majority of smart packaging systems. The distinctive feature of the latter is simple and helpful functionality which, firstly, prevents people handling the products from errors, secondly, protects the product consumers' health, and, thirdly, lowers the expenditures on the after-use package utilization.

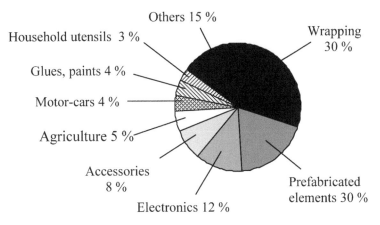

Fig 2.39. Key application areas for plastics (Prince 1998)

2.6.1. Active packages

Active packaging materials exert (upon the packaged product) influence of chemical, physical or biological nature by changing the content of gas medium inside the package or by transforming the surface layer structure of the packaged products or materials.

Inhibited polymer film contains a microporous layer which includes a complex of contact and volatile corrosion inhibitors (CI); the film is intended for anticorrosion protection of metal ware (Pinchuk and Neverov 1993). The article packed in such film is initially protected against corrosion owing to transfer of the contact inhibitor from the film to the articles surface. This protection mechanism is especially effective at vacuum packaging. As the volatile inhibitor evaporates, additional partial pressure of the second inhibitor is created in the cover; adsorption of the second inhibitor on the surface of the article prevents corrosion for a long period of time (years). Evaporation in the air-tight casing stops when the pressure critical value is reached. At the casing loss of sealing, evaporation continues creating a protective environment inside the package until the complete stock of the inhibitor in the film is consumed (Goldade et al 2005).

Basic variants of such film should be related to the following taxons:

A_1B_6 is the film which (at deformation producing the contact with the packaged product) exudes and transfer the contact CI to the product surface preventing adsorption of moisture vapors and corrosion agents inside the package; this active system has no FBS which is created by 'smart' exudation of volatile CI vapors contained in the film;

A_3B_9 is the film in which (under the change of temperature driving the moisture condensation in the packaging area) the FBS operates initiating emergence of stresses in the film porous layer and release of CI in liquid or gas phases from the pores;

A_7B_{23} and A_7B_{29} are the films where (under the action of aggressive environment) the FBS operates optimizing CI delivery to the packing by regulating concentration diffusion in the system 'film – packing space' or electrode reactions on the packed metal part.

Fig 2.40, *a* shows the microstructure of polyethylene film made by hose extrusion and thermodiffusion saturation (in the process of hose blowing) of the polymer base by the suspension of volatile CI in the oil solution of contact CI (Pinchuk et al 1991). Micropores can be seen through which the contact CI comes to the film surface and the volatile CI evaporates. Initially, the kinetic curves of G-2 evaporation from the films (*b*) represent parabolic relations $m/m_0 = a\tau^{0.5}$, where m and m_0 are the quantities of CI desorbed during time τ and $\tau_0 \to \infty$, and a is the coefficient. This is an evidence that the evaporation rate is controlled by CI diffusion in the film porous system. The kinetics of initial G-2 evaporation is linear. The volatile CI evaporation depends on its concentration in the film, its manufacturing process and increases with the temperature rise. The FBS function in this smart system is performed by the volatile inhibitor evaporation which is limited by the value of its vapors partial pressure in a sealed casing.

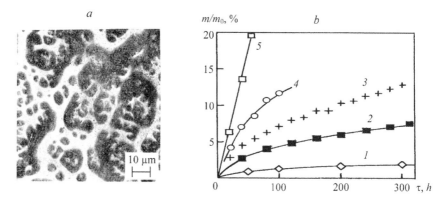

Fig 2.40. Microstructure of the film surface (*a*) and kinetic curves (*b*) of evaporation (from the surface) of the volatile CI of G-2 grade. CI concentration in the films: 1÷3 – 0.09 weight %, 4 – 1.4 weight %, 5 – G-2 in the initial state. Temperature 25 °C (*1, 4, 5*), 50 °C (*2*) and 60 °C (*3*)

A smart inhibited film may have the second operating FBS which is based on the color-change chemical reaction of the volatile CI with the filler particles in the film which change their color, for example, to green. When the inhibitor stock in the film is exhausted, the filler changes its color from green to red. This is a signal to repack the article into a casing made of a new film material (Goldade et al 2005). This film corresponds to A_7B_{26} taxon which combines smart films with the FBS based on photo-chemical reactions of the film components. The totality of such 'informa-tive' packaging materials is considered in i. 2.6.4.

A_2B_7 are the packaging films with friction coefficient regulated by the FBS changing the films surface layer structure.

Packaging films with adjustable friction coefficient f are required to solve two contradictory tasks. The first one is to increase f so that the filled film bags do not slip over each other at storing. The films made of polyethyl-ene with addition of amide compounds (4 %) manufactured by American-based Polyfil Corp. have $f \sim 0.5$ as compared to the initial value of 0.15 (Mogg 2001). The films with scabrous external layer of the baked powder particles have even higher friction coefficient (Dovgyalo and Yurkevich 1992). The opposing task is being solved in the films which demonstrate anti-friction properties at filling the film bags with loose products. This material is in demand for formation of the inner layer of the multilayer film packing.

The films in the cited examples are active materials with f = const. Evidently, 'inclusion' (into their structure) the FBS which optimizes the f value depending on the package loading modes will increase the effi-ciency of the products circulation.

A_5B_{17} are the films with the FBS which takes away the charge sponta-neously accumulated on the package in the process of its circulation.

Antistatic packaging films are convenient in operation because they decrease dust adhesion to the package and eliminate the possibility of spark discharges occurrence. The latter provides them for primarily using gunpowder and explosive powders as the material for packaging (Apparel 2001).

A_7B_{29} are the smart packaging materials in which the FBS 'oper-ates' optimizing the gas medium content inside the package by oxygen absorption

Oxygen absorbing plastics are used in foodstuffs and beverages pack-aging. Cheap technological additives on the base of iron salts and active UV-responding substances are introduced into the plastics binder. The additives must be 'friendly' to the plastics processing technology. Ube Industries, Japan, developed the manufacturing technology without addi-tives: oxygen-capturing molecular centers are engrafted to the polymer chain. These plastics can be used for manufacturing beer bottles, contain-ers for juices, sauces and dressings, films for bread packaging (Martin 1999). Materials of this kind are used in plastic corks for wines and bever-ages. The corks obtained by extrusion of expanded composition do not only seal the bottles reliably (as they are less gas-permeable as compared to the natural corks), but also help to preserve the wine taste and flavor for a long time (Tomer 1999). The FBS function in these plastics is performed by non-linear temperature- and pressure-dependent changes in absorption characteristics of absorbents.

A_9B_{29} are the materials which regulate the microbial composition of the medium inside the package by the actions of chemical nature.

Packaging films for meat products contain natural and synthetic food preservatives. Maximum storing term for chilled ($T \approx 5÷7$ °C) pork and beef packaged in the active film modified with mustard or coriander oil is 2÷3 times higher than when packed in the standard film (Goldade et al 1998). Such storage does not affect the meat organoleptic properties, its chemical and microbial compositions remain in line with the corresponding standards. Intensive development of the meat products microbial damage processes is usually accompanied by the pH abrupt jump into the alkaline area along with an increase of the volatile fatty acids content in the meat. Introduction of preservatives (ascorbic and nicotinic acid, glucose, complexes of 'salt – conjugated carboxylic acid' type) into the packaging film stabilizes the product pH at the initial stage during the long-term storage period and has favorable effects on the meat products quality (Makarevich 2000).

The generation of plastics with the antimicrobial activity has been under development since 1990s. These materials are used in the production of the following: ship plating, walls and roof coverings, boiler casings, kitchen boards, children's toys, sandals, shower curtain, stationery, waste containers, etc. (Grande 1997).

Biocidal packaging films are used mainly for medicinal, sanitary and hygiene purposes. Medical instruments, disposable razors, creams are packed in them. Biocidal adhesive films are applied onto the skin areas to be surgically dissected. Such films contain additives which suppress the growth of microbes; the films are manufactured in 'clean' rooms (Leaversuch 2001). The films with antimicrobial activity are typical active materials. They could become smart ones if equipped with the FBS which regulates the recovery of antimicrobial agents upon the signal of sensors controlling the microbial media status.

Films for non-food products protection against insect damage contain insecticides (chemical agents that deter the insect growth and kill them). The vapors of these substances are released from the film to the packaging space creating the protective atmosphere there. Use of synthetic repellents (permethrins and other chemicals of II–III hazard classes) safeguards the textiles, fur and leather products against both insects and rodents (Pinchuk et al 2001a). Introduction to the film of safe-for-humans volatile repellents on the base of herbs and essential oils repels the insects without causing damage to the consumer properties of the packaged product, health of maintenance personnel and users of the product (Goncharova et al 2009). The FBS function in such packages is performed by the ability the films to evaporate the volatile elements which is being changed naturally depending on the medium content inside the package.

So the active SM protect the packaged products against damage owing to a combination of the barrier effect with the purposeful structure change of the packaged goods surface layer and/or medium content change inside the package.

2.6.2. Gas-selective packages

The gas permeability of films is an integral characteristic of the process which includes the stages of gas adsorption on the film surface, gas diffusion in the film, particles excretion on the other side of the film and their desorption from its surface. Every stage can be accompanied by dissociation and ionization of gas molecules, chemical reactions between them and macromolecules of the film material. Gas permeability selectivity is the film property to pass different gases at different rates. Permeability factors of i and j components of gas mixture $P_i = D_i\delta_i$ and $P_j = D_j\delta_j$, where D and δ are the diffusion and solubility coefficients. The mixture separation coefficient in the film is determined by the ratio of the components permeability coefficients $F_{ij} = P_i/P_j = D_i\delta_i/ D_j\delta_j$ (Reitlinger 1974).

In the food industry the requirements to gas selectivity of the packaging films are determined by the chemical nature of the packaged product (fat content, acidity), its physical condition (liquid, paste, flakes, powders, grains, bread, etc.), the product sensitivity to moisture, oxygen, light or necessity of the product complete isolation from the environment. The selective gas permeability of the films can be adjusted by forming them of polymers with different flexibility of macromolecules, applying filling and plasticization and using multi-layer film systems. This enables it to protect the products against humidification, chilling action of the UV spectral band of the visible light, prevent evaporation of aromatic substances and product infestation with microbes, assure optimal parameters of gas exchange between the container and the environment. The most informative parameters of the gas-selective packaging film efficiency are constants of its permeability for three gases – N_2, O_2, CO_2 (Genel et al. 1976).

It follows from the above that the smart gas-selective films correspond to A_7B_{29} taxon – the environment diffusion action upon the packing stipulates reactions of its components with the film material which are performed under control of the FBS optimizing the film structure.

Film packing for cheese ripening is a typical example of the smart gas-selective films. Traditional cheese-making technology envisages careful control over cheese maturation and assurance of optimal conditions for formation of specific flavor and visual appeal. The manufacturers' aspiration to increase the production and sales volumes has come into antagonism with the labor intensity and duration of the traditional cheese-making process. In the 1980s, cheese (without forfeiting its best qualities) has passed a way up from expensive gourmet item category to a casual food product. This became possible owing to the technology of vacuum packaging of semi-finished cheese into gas-selective films assuring conditions for its ripening.

The film was first developed by Cryovac Co (USA), the world's supplier of multi-layer cheese-ripening films equipped with gas-selective layer of polyvinylidene chloride or its substitutes on the base of aromatic copolymers. The film combines low oxygen permeability [less than 400 $cm^3/(m^2 \cdot day \cdot Atm)$] to prevent mold growth, limited moisture permeability [less than 0,02 $kg/(m^2 \cdot day)$] to prevent shrinkage of goods and high permeability for CO_2 which is formed at cheese ripening [500÷2500 $cm^3/(m^2 \cdot day \cdot Atm)$]. The smart packaging allows to get rid of labor-intensive processes of removing mold from the cheese and to dramatically decrease the losses of product (Fig 2.41). The FBS function is performed by physical and chemical interaction of the film layers (up to 10) which regulates the diffusion flows in the package.

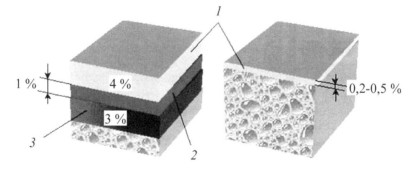

Fig 2.41. Losses of product at cheese ripening: *a* – traditional method, *б* – maturation in the gas-selective package. The figures show the cheese losses due to: *1* – drying and washing, *2* – cheese cleaning, *3* – cheese crust removal (Pinchuk et al. 2001c)

The smart gas-selective film produced in Belarus (Pinchuk et al 2005) consists of three layers. The external barrier layer limits penetration of O_2 inside the package. The layer which the cheese is in contact with (the contact layer) is made of chemically inert polymer and filled with carbohydrates – a source of energy for lactic acid bacteria. Adhesion layer used for bonding the above layers contains CO_2 absorbing substances, as well as biocides which prevent cheese from growing moldy.

The film 'works' as follows. The lactic acid bacteria 'eat up' the carbohydrate filler particles in the film contact layer at the boundary with the head of cheese. This expedites penetration of CO_2, which is produced in large amounts at the initial stage of cheese ripening, through the contact layer defects to the adhesion layer. Here, CO_2 is partially adsorbed by an adsorbent – the adhesion layer filler, while the gas excess is diffused through the barrier layer to the atmosphere. At the same time, mass transfer of biocides begins through the contact layer (which has higher permeability now) inside the package. At the main stage of cheese ripening, the biocides concentration in the package is low and does not affect the lactic-acid bacteria activity. Critical concentration of biocides on the surface of the cheese head occurs in 20÷25 days after the packaging, i.e. at the final stage of maturation when mold is formed on the cheese head. Biocides destroy mold and all cultures initiating the microbial damage of the ripened cheese (Pinchuk et al 2001c).

The FBS in this smart packaging is 'actuated' by lactic acid bacteria which open an exit of CO_2 from the package space into the environment at the initial stage of cheese ripening and delivery of biocides from the film inside the package at the final stage of its maturation.

Gas-selective films which are traditionally used for storing fruits and vegetables have constant gas permeability parameters (Genel et al 1976), thus, they correspond to the category of active, but not smart materials.

2.6.3. *Biodegradable packages*

In the middle of the 20th century, growth in the polymer production and mass use of the packaging films raised a problem of the film wastes utilization. It takes a long time (dozens of years) for synthetic polymers to decompose under natural conditions. The thrown away packaging films make up a considerable part (in various countries – from 10 to 60 %) of domestic solid wastes and have become a constant source of pollution of surface soil and coastal waters of the oceans (Ashby 2009) in the world.

Analysis of methods for polymer rubbish decontamination used in different countries shows that there is no single universal strategy to solve this problem. Two directions prevail in the technologies of polymer wastes manipulation. The first one lies in utilization (i.e. profitable use) of the processed polymer materials: multiple use of the packing (if possible), incineration or pyrolysis (high-temperature destruction in the absence of oxygen), wastes recycling along with generation of thermal energy and obtaining useful substances. These methods are characterized by technological complexity, low efficiency and high cost (Losk 1999). The second direction envisages destruction of the packaging wastes under action of the environment, e.g., by burying in the soil where microorganisms cause degradation of macromolecules. This way what causes the growth of dump areas is ecologically harmful. The urgency of the problem can be decreased by manufacturing the packages of polymer materials with controllable 'lifetime'. They preserve their initial properties during the package service life, upon expiry of which accelerated biophysical and biochemical transformations in the environment occur and the package is disintegrated. The disintegration products are included into the ecosystems metabolism processes (Netravali and Chabba 2003). These plastics are called biodegradable.

The packaging industry use biodegradable plastics on the base of the following classes of materials (Goncharova et al 2006):

- natural polymers – starch, dextrins, pectins, chitine-chitosan, cellulose, lignin, gelatin, casein, etc., which raw material resources are constantly renewed (Netravali and Chabba 2003);
- polymers which are similar in structure to biopolymers, obtained by polymerization of monomers – derivatives of nontraditional raw material base or by inoculation of the functionalized oligomers to the synthetic polymers chain (Vasnev 1997);
- high-molecular compounds which are the product of microbiological synthesis of organic raw materials and chemical products (butylene glycol, butyrolacton, butyric and chlorine-butyric acids) (Pkhakadze 1990);
- composite material – synthetic polymers filled with natural polymers (starch, cellulose, chitin) or non-organic substances containing elements which are attractive for microbes – Ca, N, K, P, etc. (Makarevich 2000);
- electret polymer materials – the sources of the constant electric field which facilitates immobilization and growth of the soil microbes in the buried film and assists in its accelerated biodegradation (Goncharova 2010).

It is necessary to clarify the following. Ideally, the biodegradable plastics are being destroyed under the action of microorganisms to simple compounds (H_2O, CO_2) which are easily assimilated into the environment. In the case of composite plastics, the microbes completely destroy biopolymers and nutrient salts. The remains of the polymer binder continue to degrade in the soil at a growing rate losing their strength and weight. The starch-filled film transforms into a perforated structure and disperse particles and then – into low-molecular fractions which pose no harm to the environment. It would be more accurate to call these plastics 'biodestructible' (Goncharova 2010).

The structure and properties of materials of this class are constant during operation and start to degrade after its completion. The extreme influence of the environment, e.g., ozone, sunlight, microbes, – serves as a signal to the end of the plastic 'lifetime'. Such natural factors are not elements of FBS. This means that biodegradable films are multifunctional materials which demonstrate (under certain conditions) additional property of accelerated biodegradation. Which features of the artificial intellect would allow them to perform their functions more success-fully? Firstly, the films shall not lose their original properties until the end of their operational lifetime by not responding to irrelevant signals. Secondly, the ideal film shall degrade immediately upon expiry of its service life and not during several months or years. Thirdly, if there is no tool to initiate the time-expired film structure collapse, it is desirable that immediately after completing the operation the film degradation occurs at a maximum rate by transforming it into small fragments and the disintegration (that may continue for a certain period of time) would not cause harm to the environment. The SM of the first two types does not exist yet. The third type can be exemplified by the electret biodegradable film.

Polarization-charge-carrying biodegradable film was developed because of the fact that chemical and technological resources of the plastics accelerated biodestruction had been practically exhausted. However, it was found out that the electret films made of thermoplastics with effective surface charge density $\sigma_{eff} = 4 \div 8$ nC/cm² stimulates immobilization of the soil microbes, growth of their colonies and accelerated biodestruction of macromolecules. Strength of electret films in the soil decreases for 60 % and weight – for 40 % in the course of year (Goldade et al. 2011).

Fig 2.42 shows the kinetic curves of the change in strength of starch-filled polyethylene (PE) films at exposure in the soil.

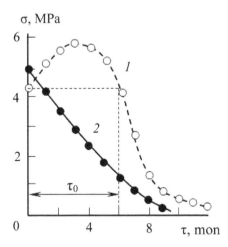

Fig 2.42. Decrease in the breaking strength at tension of PE films vs exposure time in the soil: *1* – non-electret, *2* – electret film with $\sigma_{eff} = 5$ nC/cm² (Goldade et al. 2011)

Initially, strength of non-electret films (curve *1*) increases due to migra-tion of PE stabilizers and active chemical action (upon PE) of humic acids available in the soil which initiate the secondary PE crystallization and macromolecules cross-linking. Then the strength exponentially decreases in proportion to the increase of size and amount of microbial damages. In 9÷10 months after burial in the soil, the film represents a fragile perforated structure with the strength approaching zero. Thereupon, the task set at creation of biodestructible films may be considered as solved, as such structure which is water and air permeable, does not pose any danger to breathers and does not affect the ecological balance. Film fragments disintegration to end products in the form of H_2O and CO_2 are the final stages of the film biodestruction.

The kinetics of the electret films biological destruction is considerably different (curve 2). The initial strength of electret films is slightly higher than non-electret ones, as the polarization charge formation is accompanied by the structure ordering of polymer binder. Loss of strength starts immediately after the film is buried in the soil. This occurs due to activating influence of the film's own electric field onto the kinetics of the adsorption immobilization of microorganisms, intensity of their metabolism, onto the growth rate and activity of the microorganisms colonies grown on the film. This is not to say that the aging processes which stipulate its structure rearrangement cease to exist in the film. However, the competing process of macromolecules demolition by microorganisms-destructors occurs more intensively than the process of the film strengthening. As the result, the period of film biodestruction in the soil is shortened by τ_0 time in the course of which strengthening of non-electret films with subsequent reduction in their strength to the initial value occurs. The electret film demonstrates features of the SM in which the FBS function is performed by the electret field. Initially, it attracts microorganisms and after their immobilization it decreases exponentially since the charge carriers in the film 'being eaten' by microbes serves as the source of energy for them. The smart electret biodegradable film corresponds to A_9B_{18} taxon.

In conclusion it should be noted that the biodegradable films are favorable domain for creating smart systems at the junction of animate and inanimate nature. The FBS functions in them can be performed by metabolism processes which are naturally inherent in the living matter.

2.6.4. *Informing packages*

The informing package tells the product distributor, seller and consumer the operative data on the quantity, location and condition of the packaged product – primarily, on the freshness of the foodstuff, temperature, medium composition inside the package, etc. At the present time, the most comprehensive information on the packaged product is provided by RFID (radio frequency identification) technology. RFID device consists of a microchip measuring less than 1 mm in length connected to an antenna which transmits the information when entering the RFID scanner field. The reading distance of the tag with radiation, e.g., 13.56 MHz is a quarter of a meter, while 915 MHz tag can be 'seen' at a distance of three meters. RFID technology is not expensive and is highly effective to control the packaged product (Teplyakov 2006).

The RFID-transmitted data are based on signals coming from the package elements which exercise the functions of sensors – thermochromic, photoelectrochemical, liquid-crystal, etc. Of course, smart packaging films fall short in accuracy and information completeness as compared to traditional sensors (thermistors, photodetectors, digital cameras, etc.) but surpass them in multifunctionality and technological effectiveness of use in packaging. Modern trend in the materials science called 'sensor revolution' is implemented in the RFID system (Byrne et al. 2010). Smart informing films are the result of this revolution. It is impossible to describe all types of informing packaging films in one paragraph; their number is increasingly growing over time at an exponential rate. The most typical examples of these films are cited below.

A_3B_{16} are the smart films which respond to the temperature changes by changing their color, i.e. by changing the reflected light wave intensity or length.

Thermochromic films contain fillers – thermochromes (NaO_2, KO_2, RbO_2, CsO_2 crystals). When temperature rises, decomposition of metal oxide occurs: $MO_2 \rightleftarrows M_2O_2 + O_2$, be attended by change of color (Barukhin and Babkin 1995). The FBS is inherent in the very nature of thermochromes structure: color change corresponds to the temperature change.

A_8B_{13} are the films whose structure is changed under chemical influence of the environment and is controlled by the FBS transforming the structural changes into the changes of film optical characteristics.

Humidity indicator films consist of the lower colored and upper antireflective layers. The latter is made of the copolymer of ethylene and vinyl acetate filled with silica gel. It becomes transparent in the humid atmosphere and renders the color of painted layer (Barukhin and Babkin 1995).

A_8B_{16} are the smart films which enter into chemical reactions with the substances released by the packaged product during storage, under the control of FBS changing the film optical parameters.

The freshness indicator films for products packaging change their color under the action of gases produced at the product spoilage. The films are filled with the indicator substances which color and luminescence change at interaction with gases by the following mechanisms: acid-base (reaction to H^+ ions), adsorption, oxidation-reduction (reaction to the change in the oxidation-reduction potential), complexometric (reaction to metal ions) (Simonova 1998). Packaging color-changing films have been developed that give signal on milk turning sour, meat and fish microbial spoilage, etc. The FBS function in these packages is performed by the indicators sensitivity to the active gas components concentration in the package.

The smart category includes a large number of films which are not packaging ones. They are: color film dosimeters of ionizing radiation (A_4B_{16} taxon) (Barukhin and Babkin 1995); multilayer films containing a liquid crystal layer (A_6B_{13}) used in magnetotherapy (Sonin 1983); photo-electric organic films (A_5B_{16}) for flexible displays (Jang 2006), as well as radiation dosimeters in the form of film strips subject to shrinkage under the action of radiation proportionate to the absorbed dose (A_4B_3) (Selkin 2004).

In conclusion it should be noted that the packaging industry is one of the most 'advanced' branches of production in terms of introducing the artificial intellect features into its products. RFID system development provides for keeping the data on all packaged products in the Internet with no limitation on time and space. Negative reactions to that by some users is caused by an opinion that provision of information through RFID may become an instrument of surveillance and intervention into the private life (Teplyakov 2006). This assumption was used as a reason for instituting legal proceedings on permissible limits of RFID usage. Nevertheless, the decision on RFID technology application for goods marking was adopted by the European Union in 2004.

2.7.   Biologically active materials and systems

The biologically active materials which change the structure and properties of the biological environment are all materials used as implants (described in i.2.5.3), means of target-oriented drug delivery (i.2.5.4) and scaffolds for growing biological tissues (i.2.5.5). Information on other SM included into the multitude of bioactive materials is given below.

2.7.1.   Antimicrobial materials

The data on certain types of polymer SM suppressing the growth of pathogenic microorganism is given in i.2.6.1 (plastics regulating microflora in the packaging space) and in i.2.5.2 (adhesive antiseptic films applied onto the skin areas to be surgically dissected).

Biocides-containing plastics are used to cover handrails in the public transport, tables in public spaces (post offices and baggage services, kitchen furniture in public catering facilities), in medical institutions (first-aid and medical treatment rooms, operating rooms). There is a wide range of products which can reasonably be made of plastics modified with antimicrobial substances: trays, chopping boards and containers for foodstuff, children's toys, office appliances (phones, keyboards, mouses), toilet bowls, sinks and other sanitary ware. Coatings on the base of biocides-containing varnishes and paints are applied to prevent biological fouling of ships (periphyton). The periphyton base consists of bacterial slime which plants and animals (algae, crustaceans, mussels, hydroids, sponges, etc) are attached to.

The aforementioned polymer materials are included into A_9B_{29} taxon – materials which respond to microbial 'attack' by chemical 'attack' on the immobilized microorganisms. The response of active materials without FBS to biological 'colonization' of the products surface is characterized by constant intensity. This stipulates inefficient consumption of antiseptic components and unjustified reduction of the engineering resource of the active material as an antimicrobial agent. Plastic coating with the surface layer in a jelly-like state exemplifies SM of this class. The polymer matrix is supplied by a system of interconnected micropores filled with antimicrobial liquid. The latter released through the micropores to the coating surface by syneresis mechanism (spontaneous release of liquid) (Goldade et al 2005). Syneresis forms the basis of FBS; it enhances at temperature rise when the activity of immobilized microorganisms increases and terminates at low temperatures wherein the metabolic processes practically come to a stop.

Antimicrobial synthetic fibers are used in new generation of textile products whose protective properties are adapted to the modern human habitat. The smart clothes are characterized by multifinctionality: ideally they should protect humans not only against pathogenic microbes, but also against ultraviolet light, electromagnetic waves, static electricity, etc.

The first antimicrobial synthetic fibers have been developed in military clothing. The modern military forces in China wear antimicrobial clothing, in England there is a governmental program '21st Century Warrior' which envisages development of antimicrobial clothing. In 2005 the demand for antimicrobial fibers in the Western Europe amounted to 17.6 thousand tons, in 2011 that parameter exceeded 30 thousand tons. For 20 years already, the world leading textile companies have been manufacturing antimicrobial chemical fibers which, most frequently, organic antibacterial agent Triclosan or silver compounds are introduced into. These modifiers suppress the microbial growth and in microamounts are safe for humans. Antimicrobial synthetic fibers can be added to any fabric – on the base of cotton, lint, silk, wool. They make the fabric crease-resistant and wear-proof and add to it antiseptic properties.

The problem which the developers of such fibers are facing is the long-term preservation of the antimicrobial effect. Its solution is complicated by the fact that textile products are periodically laundered; due to this process modifiers which have been introduced into the fibers structure by traditional methods (impregnation, binder filling) are removed quickly from the surface layer of the fibers.

One of the ways to solve the problem is to introduce antimicrobial substances into synthetic fibers by the mechanism of crazing. Crazing is the phenomenon which occurs at the fibers drawing in surface-active liquids. A system of specific microcracks (crazes) develops in the extended fibers; the walls of crazes are connected with fibrils (Fig 2.43). Colloidal particles of antimicrobial substance available in the surface-active liquids penetrate the crazes and are adsorbed on the newly formed walls. After the fiber is unloaded, the crazes close and the captured microamounts of active substance remain protected (in the surface layer of the fiber) against external influence (Pinchuk and Goldade 2014). At the same time they can be released to the fiber surface at deformation of the fibers, humidity growth, temperature rise. This specific feature of synthetic fibers with antimicrobial substances introduced into the crazes perform the FBS functions and imparts to the fibers the SM features corresponding to A_9B_{29} taxon.

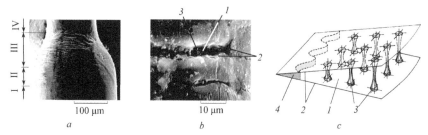

Fig 2.43. SEM images (*a* and *b*) and diagram (*c*) of crazes formation in polyester fibers. Roman numerals indicate the areas of the fiber tensioned state where the following occurs: I – crazes initiation, II – growth, III – crazes spreading, IV – formation of the fiber fibrillar structure. *1* – walls of crack, *2*– crack, *3* – fibrils (Pinchuk and Goldade 2014)

2.7.2. Regulators of fermentation

Fermentation is the enzymatic degradation of organic substances (mainly carbohydrates) which is evolutionarily earlier and energetically less advantageous form of energy retrieval from organic raw materials as compared to the ordinary oxidation. It may occur in the organisms of animals, in plants and microorganisms with or without involvement of O_2 (aerobic or anaerobic fermentation). Fermentation is a source of energy for microbes activity (some of them grow only owing to fermentation energy) while anaerobic degradation of organic wastes (primarily, cellulose ones) plays an important role in the circulation of elements in the nature. Fermentation is used at production of various engineering and food products. Spirit fermentation is used in wine and brewing industry, fuel generation; lactic fermentation – at production of fermented milk products and lactic acid, at feeding-stuffs silage; propionic acid fermentation – in cheese-making industry; acetone-butyl fermentation – for making solvents, etc. In the industrial sector the fermentation processes are performed by methods of biotechnology.

$A_9B_{31,\,32}$ are the smart biosystems which process the organic substances under control of the FBS regulating the processes of biochemical fission and synthesis.

Immobilized enzymes are an achievement of the modern biotechnology. They are artificially generated preparations of enzymes which molecules are physically and chemically bonded to the carrier (usually a polymer one) preserving (completely or partially) its catalytic properties. Activity of the immobilized enzymes is $30\div90$ % of the source enzyme activity. Enzymes immobilized on the carrier with the help of covalent, ionic or coordination bonds have maximum catalytic activity (Perepelkin 1998). During immobilization the conditions are created to prevent blocking of the enzyme catalytic center by the carrier's active groups and to exclude loss of its activity.

The immobilized enzymes are a convenient engineering tool for fermentation control. They are introduced into the organic environment to initiate fermentation or remove from it to stop fermentation at any stage. External and internal diffusion barriers between the medium and the immobilized enzyme have a significant influence on the fermentation kinetics. The first one is formed by the unstirred thin layer of the medium around the enzyme particle (Nernst's layer). The internal diffusion barrier emerges due to limitation of the medium diffusion inside the polymer network which carries the enzyme particle (Kozlov 1990). Advantages of such fermentation regulator are, firstly, the increased stability toward the denaturing effects (heating, aggressive media influence, autolysis, i.e. cells self-dissolution, etc.) and, secondly, possibility of complete separation of the immobilized cells from the fermentation medium without polluting the resulting product with the microbial mass.

The immobilized enzymes are used as fermentation regulators: at obtaining amino acids, antibiotics (penicillin, prednisalon), chemical reagents via method of fine organic synthesis; for removal of endotoxins formed in wounds and skin burns; in immunoenzymometric laboratory and clinical methods of analysis; at production of food and beverages, including the lactose-removed ones for people with lactose deficiency.

The majority of the immobilized ferments are typical active materials. In the fermentation systems where these materials are used, the FBS function is performed by a technologist who regulates the time of enzymes contact with the working environment. The smart 'enzyme – carrier' systems shall be equipped with the FBS which regulates the fermentation process according to the required degree of the source product fission. It optimizes, firstly, the rigidity of the enzyme particles binding to the carrier, and, consequently, activity of the immobilized enzyme, and, secondly, the height of diffusion barriers between the product and the enzyme particle. Chemical engineering has not created such fermentation regulators yet.

2.7.3. *Regulators of plants growth*

Plants growth is the irreversible increase of the size and mass due to new formation of the plants structure elements – cells, tissues and organs. Basic processes which determine this phenomenon – division of cells, their growth and differentiation (occurrence of differences between homogeneous cells) – take place in the specialized tissues (meristems). Four phases are differentiated in the kinetic dependencies of the plants growth: initial, logarithmic or lag phase of intensive growth, growth retardation phase and steady state condition. The natural growth of plants occurs rhythmically, i.e. the phases of intensive and retarded growth are interchanging. The rhythms follow the changes in the environment or are controlled by genetic inner factors. A group of substances has been discovered – growth hormones or phytohormones, some of which stimulate and others inhibit the plants growth (Wareing and Phillips 1981).

The materials – regulators of growth – 'switch on' or 'turn off' the mechanisms determining the plants growth at various stages of development. It would be interesting to examine the regulators of growth of all species known in the world – humans, animals, plants and microorganisms. We have limited ourselves to the world of plants, firstly, keeping in mind the engineering subject matter of the book, secondly, because of the fact that specific aspects of metabolism regulation in the organisms of humans, animals and microorganisms have been already discussed, and, thirdly, understanding that regulation of growth in all species of the animal world is a huge subject and it is impossible to cover all the details in a single book.

A_4B_{32} are the smart materials which transform the visible light into radiation stimulating the plants growth under the FBS control which regulates the luminescence intensity.

Photocorrecting films are the materials used for covering hothouses and greenhouses in agricultural industry. They contain luminophors transforming the UV constituent of the natural or artificial light into the radiation of the read area of the spectrum. This facilitates the increased growth (at the initial and lag phases) of agricultural crops (Minich and Raida 1998). Short-wave blue-violet light in the opposite area of the spectrum produces the same effect on photosynthesis (Vinchester 1967). Functional capability of the films – growth photocorrectors is determined by two parameters: luminescence intensity I (a characteristic of the material ability to transform UV radiation into the red light) and service life t of the luminophor in the film (luminescence duration, technological lifespan of the film). Creation in the photocorrecting film of the FBS which regulates *I* depending on the plant growth phase will provide for increasing efficiency and *t* of the smart film.

A_6B_{32} are SM which magnetic field stimulates the plants growth under the control of FBS adjusting the field intensity depending on the growth phase.

Magnetobiology is a division of the biophysics which subject matter is to study the influence of external magnetic fields on living organisms. The most impressive results of the magnetobiology have been received in the non-gravity environment, although magnetic stimulation of the plants growth has been known since ancient times. We think that the main achievements of the magnetobiology will be attributed to creation of the smart magnetic stimulators of plant growth which application will not complicate the agricultural engineering.

A_7B_{32} are the smart films and coatings which regulate the development of plants root system, fruits and seeds by optimizing the content of their environment under the control of FBS of diffusive nature.

Polymer coatings of the root system of the forest seedlings are meant for solving the following tasks: protection of the roots against mechanical damage; moisture conservation; feeding the root system with nutrient elements during storage, transportation to the planting site and seedlings engraftment. These tasks are solved by coatings formed of aqueous solutions of polyacrylamide, sodium carboxymethyl cellulose, urea-formaldehyde resins, etc. The smart coatings create (around the root system) the microatmosphere with optimal humidity and nutrients content. The FBS function in the 'coating – roots' system is performed by the natural aging-caused transformation of the structure of coatings with optimal composition during the seedlings engraftment. Coatings biodegradation stipulates favorable (for the root system) changes in their elasticity, relationship between water permeability and water absorption, pH of the root system surrounding medium. The accelerated engraftment and growth of seedlings have environmental impact (production of oxygen, improvement of climatic and sanitary-hygienic conditions in the region, protection of waters, etc.) which exceeds (in terms of money) the cost of standing wood resources more than threefold (Kopytkov 2007).

Mulching is the covering of the soil (complete or in spaces between rows) with paper, humus, and the like. It reduces water evaporation, decreases amplitude of daily temperature fluctuations in the soil, prevents formation of the soil crust. Evidently, the structure of smart mulch films shall be microporous. The micropores cross-section is regulated by the FBS and is changed during the day due to thermal expansion and covering the pores with water condensate droplets, while during a longer time period – due to biological damage of the film by soil microorganisms.

Films for fruit ripening optimize gas exchange between the environment and the membraneous capacity which the fruits are placed into. Owing to that, the morphological, biochemical and physiological changes are accelerated in the fruits, as the result of which proteins, fats and carbohydrates are intensively formed, and the seeds contained in the fruit become full-fledged ovules of new plants. Ethylene is accumulating in the ripening fruits; it suppresses biosynthesis of auxins (plant hormones that slow down fruit ripening). The process of respiration, which furnishes energy to the fruit tissue, intensifies.

The smart film for fruits ripening, firstly, must be gas-selective, i.e. it shall pass both O_2 which is intensively absorbed by the ripening fruits (mainly from the environment into the packing) and CO_2 which is released by the fruits at ripening (in the reverse direction, i.e. from the packing into the environment). Secondly, the free space of the film shall be saturated with ethylene. Although such films have not been created yet, the modern chemical technology has various tools sufficient for solving these tasks.

A_8B_{32} and A_9B_{32} are the smart systems which regulate the plants growth by chemical and biochemical actions under the control of FBS optimizing the biosynthesis intensity in the plants.

Phitohormones are the organic substances secreted by specialized tissues of higher plants. In negligibly small amounts they act as the regulators and coordinators of ontogenesis – individual development of a specimen (Wareing and Phillips 1981). Phitohormones concentration in cells is changed instantly at the attack of any pathogen. Thus, an increase in endogenous content of phitohormones provides for mobilization of reserve defense mechanisms of plants. *Herbicides* are chemical preparations used to kill unwanted vegetation. Selective herbicides are the smart herbicides which kill weeds without damaging the crops. In view of the environmental hazard of uncontrolled application of herbicides, their use is regulated by law in many countries. Development is expected of the smart biotechnical materials, primarily, films modified with phitohormones or herbicides. The FBS must regulate their release from the film under the signal of sensors responding to, firstly, the attack of pathogens and, secondly, to the plants 'respiration' which is different for weeds and crops.

Hydroponics is a method of growing plants without soil. Water with the dissolved nutrient elements is filtered to the roots bound in gravel, moss, sand medium or sprayed onto the roots which are located in air. The FBS function in this system is performed by the natural ability of the plants to process the nutrient substances in the amount corresponding to the genetically predetermined metabolic capabilities.

2.7.4. Filtering materials for water cleaning

The modern trend of the industrial production which consumes huge quantities of water lies in the increasing role of its biocleaning systems. The utility for wastewater biological cleaning – biofilter – represents a reservoir filled with coarse-grained filtering material (carrier) which microorganisms colonies are adsorptionally bound to (immobilized). When passing through the filter, the wastewater is mechanically cleaned. Its organic contaminants are settled on the filtering material grains and are used as nutrients and source of energy for microorganisms. Thus, the biofilter is a smart biosystem which removes admixtures from the water and cleanses itself owing to the FBS whose functions are performed by metabolism of immobilized microorganisms ($A_8B_{31,32}$ taxons). Microbiological utilization of admixtures found in the wastewater cleaned through biofilters simulates the natural biochemical phenomenon of 'self-cleaning' in nature. The biofilter typical structure is shown in Fig 2.44.

The main factor and the key link in the trophic chain of admixtures bioutilization is availability of active strains of microorganisms and their consortiums. The efficiency and reliability of biofilter operations are determined by multifunctionality of the biomass carrier to a considerable extent. It must ensure the required level of water cleaning from admixtures, be sufficiently water- and gas-permeable, retain the biomass, protect it against mechanical, aero- and hydrodynamic overloads. From this standpoint, the most significant parameters of the carrier are the following: porosity, developed surface, satisfactory mechanical strength and chemical stability, biocompatibility.

Fibrous polymer structural elements obtained by depositing (with gas flow) the fibers of polymer melt onto the shape-generating scaffolds (*melt blowing* technology) possess these properties. For industrial biofilters, the optimal carriers (according to the afore-mentioned criteria) are rings made of polypropylene or polyamide fibers welded in the points of contact. Rings dimensions are as follows: mean diameter 40 mm, height 30 mm, wall thickness 5 mm. Fibers diameter is 50÷100 mcm, fiber material density is 200 ÷ 250 kg/m³ (Pinchuk et al 2002). Advantages of such carriers are in their low bulk mass (100 ÷ 120 kg/m³), chemical inertness and wide technological possibilities to add extra-functional properties (sorptional, electret, magnetic) to the carriers' fibrous base. Firstly, it increases the water purification efficiency at the expense of implementing several mechanisms for admixtures capture by the biofilter. Secondly, new properties of the carriers increase the microorganisms immobilization and promote their activity (Korotkii 2005).

The distinctive aspects of melt-blown carriers which impart the SM properties lie in enhancement of the fibers immobilization activity owing

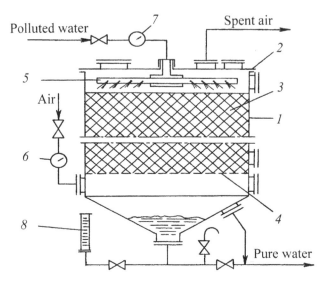

Fig 2.44. Diagram of a biofilter for aerobic water treatment. *1* – casing, *2* – cover, *3* – layer of the carrier with the immobilized biomass, *4* – supporting grid, *5* – sprayer unit, *6* and *7* – flow meters, *8* – level indicator (Pinchuk et al 2002)

to formation of technological electret charge on them and introduction of biogenic elements.

In the *melt blowing* process, the fibers acquire spontaneous electret charge with density of about 0.1 nC/cm^2 whose value depends on the process technological behavior. The charge develops in the process of the melt fibers formation and transportation due to localization of electrons on the easily polarized segments of macromolecules with unsaturated bonds (Pinchuk et al 1998). At filling the binder with ferrites, the fibrous carrier of biomass becomes a source of constant magnetic and electric fields (Pinchuk et al 2002). Filling or surface modification of fibers with non-organic salts – $(NH_4)_2SO_4$, KH_2PO_4, etc. – which contain nutrients that are used by microorganisms expedites their immobilization (Korotkii et al 2004). As a result, the carriers acquire the SM properties where the FBS function is performed by microorganisms. This occurs in the following manner. Weak electromagnetic fields and biogenic additives attract microorganisms which populate the carriers actively. Due to immobilization, the electret carrier field is weakening (Pinchuk et al 2009) and the nutrients concentration on the fibers is decreasing. Thus, transition of the immobilized microorganisms to a new source of nourishment – organic admixtures captured by the biofilter – is initiated.

The range of bioactive SM includes smart hydrogels which (being chemically and biochemically bound with the enzymes, antibodies, growth factors, receptors) perform molecular 'identification' of foreign substances in the organism by responding to them with the transition from expansion to collapse or with gel-sol transition (Ulijn et al 2007). Bionanoelectronics materials are the newest category of bioactive SM. In these systems, interaction of electronic and biological components which differ markedly from one another in nature and information mechanisms of signal transmission is implemented. Bionanoelectronics is a new paradigm for using the metabolism principles in bioelectronic devices. It is expected that the bionanoelectronics development will revolutionize many methods of medical diagnostics, endoprosthesis replacement, as well as basic 'man – computer' interactions (Noy et al 2009).

References

Adintsova, Z.N., V.Ya. Prushak and A.V. Protosenya. 1996. Fibrous Noise-insulating Materials. Infotribo, Gomel.

Agbenyega, J. 2011. Electrospinning has nanofibers in alignment. Materials Today 11/12: 10.

Alekseev, A.G., O.M. Guseva and V.S. Semichev. 1998. *Composite Ferroplastics and Electromagnetic Safety*. St.-Petersburg State University, St.-Petersburg.

Aleshin, A.N., B.S. Bokshtein and L.S. Shvindlerman. 1982. Research of diffusion on individual grains boundaries in metals. Surface. Physics, Chemistry, Mechanics 6: 1–12.

Alitalo, P. and S. Tretyakov. 2009. Electromagnetic cloaking with metamaterials. Materials Today 12/3: 22–29.

Andreev, V.S., L.P. Gavryuchenkova and V.G. Popov. 1989. Electropolarizing cleaning of suspensions of bacteria and viruses on sorbent 'Segnetal'. Biotechnology 5/4: 479–484.

Andreeva, D.V., D. Fix, H. Mohwald and D.G. Shchukin. 2008. Self-healing anti-corrosion coatings based on pH-sensitive polyelectrolyte/inhibitor sandwich-like nanostructures. Advanced Materials 20/14: 2789–2794

Apparel, L. 2001. Sparks conducted with special agents. Modern Plastics Int. 31/8: 63.

Arkharov, A.M. and L.D. Kharitonova. 1978. Friction under low temperatures. pp. 361–375. *In*: I.V. Kragelskii and V.V. Alisin (eds.). *Friction, Wear, Lubrication*. V. 1. Mechanical Engineering, Moscow.

Ashby, M.F. 2009. *Materials and the Environment: Eco-Informed Material Choice*. Elsevier. Butterworth-Heinemann, Berlin.

Asnis, L.A., V.P. Vasil'ev, V.B. Volkonskii and H.V. Hinrikus. 1995. Laser Distance-Measuring. Radio and Communication, Moscow.

Avrushchenko, V.N. 1978. Rubber Sealants. Chemistry, Leningrad.

Babkin, V.T., A.A. Zaichenko, V.V. Aleksandrov, B.F. Byzalov, V.N. Ivanov and D.P. Yurchenko. 1977. Tightness of Static Connections of Hydraulic Systems. Mechanical Engineering, Moscow.

Baklitskaya, O. 2007. Nobel Prizes of 2007. Giant magnetoresistance – triumph of fundamental science. Science and Life 11: 14–18.

Balazs, A.C. 2007. Modelling self-healing materials. Materials Today 10/9: 18–23.

Balkevich, V.L. 1984. *Technical Ceramics*. Strojizdat, Moscow.

Barachevskii, V.A. 1998. Photochromic materials. pp. 363–364. *In*: A.M. Prokhorov (ed.). *The Physical Encyclopedia, Vol. 5*. Big Russian Encyclopedia, Moscow.

Barukhin, S.B. and I.Yu. Babkin. 1995. Composite materials of new generation. Chemistry of High Energy 29/2: 126–132.

Basov, N.G., V.B. Rozanov and N.M. Sobolevskii. 1975. Laser thermonuclear synthesis in energetic of the future. Proceedings of the USSR Academy of Sciences: Energetic and Transport 6: 3–21.

Batygin, V.N. and V.G. Bravinskii. 1995. Radioabsorbing and radiotransparent materials. pp. 332–334. *In*: N.S. Zefirov (ed.). *The Chemical Encyclopedia, Vol. 4*. Big Russian Encyclopedia, Moscow.

Belov, K.P. 1998. Magnetostrictive phenomena. Materials with giant magnetostriction. Soros Educational Journal 3: 112–117

Belyaev, V.I. (ed.). 1988. *Theoretical Bases of Surface Plastic deformation*. Science and engineering, Minsk.

Belyi, V.A. and L.S. Pinchuk. 1980. *Introduction to Material Science of Sealing Systems*. Science and Engineering, Minsk.

Belyi, V.A., V.A. Goldade, A.S. Neverov and L.S. Pinchuk. 1980. Inductive metal-liquid seal. USSR Inventor's certificate # 709,878.

Bishard, E.G. 1988. Thermomagnetic materials. pp. 114–121. *In*: Yu.V. Koritskii, V.V. Pasynkov and B.M. Tareev (eds.). *The Directory on Electroengineering Materials, Vol. 3*. Energoatomizdat, Leningrad.

Black, J. 1994. Tissue response to exogenous electromagnetic signals. Orthop. Clin. North Am. 15: 15–31.

Bland, S. 2011. Higher and higher. Materials Today 14/4: 13.

Bocharov, A.M., P.V. Sysoev and V.S. Mironov. 1984. *Process of crystals processing*. USSR Author's certificate # 1,104,769.

Bogolyubov, V.M. and G.N. Ponomarenko. 1999. *General Physiotherapy: Textbook*. Medicine, Moscow.

Bohner, M. 2010. Resorbable biomaterials as bone graft substitutes. Materials Today 1-2: 24–30.

Borisov, L.A. 1990. Sound-absorbing materials. pp. 327–328. *In*: I.L. Knunyanc (ed.). *The Chemical Encyclopedia, Vol. 2*. Soviet Encyclopedia, Moscow.

Boskolo, F.R. 2006. Liquid rubber: a clever material for high-altitude construction. Building Materials. The Equipment and Technologies of XXI century 7: 20–21.

Bradley, D. 2010. A safe reaction. Materials Today 13/5: 8.

Bradley, D. 2011. Visibly invisible optical materials. Materials Today 14/3: 65

Braunovich, M., V.V. Konchits and N.K Myshkin. 2007. *Electrical Contacts: Fundamental, Application and Technology*. CRC Press, New York.

Brittberg, M., A. Lindahl, A. Nilsson, C. Ohlsson and L. Peterson. 1994. Treatment of deep cartilage defects in the knee with autologous chondrocyte transplantation. N. Eng. J. Med. 331: 889–895.

Broeze, J.E. and W.J. Laubendorfer. 1966. Low friction bearing. U.S. Patent # 3,239,283.

Bunin, V.D. and A.G. Voloshin. 1996. Determination of cell structures, electrophysical parameters and cell populations heterogeneity. J. Colloid & Interface Sci. 180: 122 – 126.

Bushe, N.A. 1993. Estimation of metal materials role in tribosystems compatibility. pp. 130–148. *In: Tribology: Researches and Applications. Experience of the USA and CIS Countries*. Mechanical Engineering, Moscow; Allerton Press, New York

Butter, P. 2006. The whole package. Materials Today 9/4: 64.

Byrne, R., F. Benito-Lopez and D. Diamond. 2010. Materials science and the sensor revolution. Materials Today 13/7-8: 16–23.

Caloz, Ch. 2009. Perspectives on EM metamaterials. Materials Today 12/3: 12–20.

Campbell, A.A. 2003. Bioceramics for implant coatings. Materials Today 6/11: 26–30.

Capolino, F. (ed.). 2009. *Handbook of Artificial Materials*. CRC Press, New York.

Chernyakova, Yu.M. and L.S. Pinchuk. 2007. The synovial joint as an 'intelligent' friction unit. J. Friction and Wear 28/4: 389–394

Chernyakova, Yu.M., L.S. Pinchuk and L.S. Lobanovskii. 2011. Structure and magnetic susceptibility of lubricating fluids in synovial joints. J.of Friction and Wear 32/1: 54–60

Chichinadze, A.V. (ed.). 1980. *Polymers in Friction Units of Machines and Devices: Reference Book*. Mechanical Engineering, Moscow.

Chichinadze, A.V. and Braun E.D. 1979. Frictional devices. pp. 230–257. *In:* I.V. Kragelskii and V.V. Alisin (eds.) *Friction, Wear and Lubrication: Reference Book*, V. 2. Mechanical Engineering, Moscow.

Cordelia, S. 2004. Needles on the microscale cause less pain. Materials Today 7/1: 6.

Could, P. 2003. Self-help for ailing structure. Materials Today 6/6: 44–49.

Dee, K.C., D. Puleo and R. Bizios. 2000. Engineering of materials for biomedical applications. Materials Today 3/1: 7–10.

Denisyuk, Yu.N. 1981. Static and dynamic volumetric holograms. J. Experimental and Theoretical Physics 51: 1648–1660.

Dianov, E.M. and A.M. Prokhorov. 1990. Optical communication on the base of nonlinear phenomena in fiber-light guide. Herald of the USSR Academy of Sciences 10: 42–47.

Dovgyalo, V.A. and O.R. Yurkevich. 1992. *Composite Materials and Coatings on the Base of Disperse Polymers: Technological Processes*. Science and Engineering, Minsk.

Drozdov Yu.N. 1979. Transfer mechanisms. pp. 113–147. *In:* I.V. Kragelskii and V.V. Alisin (eds.). *Friction, Wear and Lubricatio: Reference Book, V. 2*. Mechanical engineering, Moscow.

Drozdov, Yu.N., E.G. Yudin and A.I. Belov. 2010. *Applied Tribology (Friction, Wear, Lubrication)*. Eco-Press, Moscow.

Dukhovskoi, E.A., A.N. Ponomarev, A.A. Silin and V.L. Telroze. 1969. Effect of abnormal low friction in vacuum at bombardment of polythene by streams of fast atoms and molecules of some elements. Reports of the USSR Academy of Sciences 189/6: 1211–1214.

Elefthereades, G.V. 2009. EM transformation – line metamaterials. Materials Today 12/3: 30–41.

Feast, M.J., S. Munro and R.W. Richards (eds.) 1993. *Polymer Surfaces and Interfaces*. John Willey & Sons, New York.

Fedorchenko, I.M., V.M. Kryachek and I.I. Panaioty. 1976. *Modern Frictional Materials*. Naukova Dumka, Kiev.

Fernandes, R. and D.H. Gracias. 2009. Toward a miniaturized mechanical surgeon. Materials Today 12/10: 14–20.

Fertman, V.E. 1988. *Magnetic Liquids: Handbook*. Higher School, Minsk.

Filippova, O.E. 2005. Smart polymeric hydrogels. The Nature 8: 11–17.

Fontenot, R.S., K.N. Bhat, W.A. Hollerman and M.D. Aggarval. 2011. Triboluminescent materials for smart sensors. Materials Today 14/6: 292–293.

Garkunov, D.N. 1985. *Tribotechnique*. Mechanical Engineering, Moscow.

Gauer, D. 1989. Optical Systems of Communication. Radio and Communication, Moscow.

Geesink, R.G., K. de Groot and C.P. Klein. 1988. Bonding of bone to apatite-coated implants. J. Bone Joint Surg. 70: 17–22.

Genel, S.V., Ya.G. Muravin and O.N. Belyatskaya. 1976. Application of film materials for foodstuff packing and storages. pp. 44–60. *In*: V.E. Gul (ed.). *Polymeric Film Materials*. Chemistry, Moscow.

Gergel, V. 1988. Solid-state diodes. p. 628. *In*: A.M. Prokhorov (ed.). *The Physical Encyclopedia, Vol. 4*. Soviet Encyclopedia, Moscow.

Ghosh, S.K. (ed.). 2008. *Self-Healing Materials: Fundamentals, Design Strategies and Applications*. Wiley, New York.

Goldade, V.A. and L.S. Pinchuk. 2009. *Physics of the Condensed State*. Belarus science, Minsk.

Goldade, V.A., A.S. Neverov and L.S. Pinchuk. 1981. The strain gauge. USSR Inventor's certificate # 870,990.

Goldade, V.A., L.S. Pinchuk and A.V. Makarevich. 1998. New polymeric packaging film for food products. pp. 234–242. *In*: M.C. Dordi (ed.). Modern Book Packaging. Indian Inst. of Packaging, Mumbai.

Goldade, V.A., L.S. Pinchuk, A.V. Makarevich and V.N. Kestelman. 2005. *Plastics for Corrosion Inhibitions*. Springer-Verlag, Berlin.

Goldade, V.A., L.S. Pinchuk, O.A. Ermolovich, E.P. Goncharova and V.E. Sytsko. 2011. Formation and biodegradation of polyethylene-based electret films. Intern. Polymer Processing 26/2: 205–211.

Golubev, A.I. and L.A. Kondakova (eds.). 1986. *Seals and Sealing Technics: Reference Book*. Mechanical Engineering, Moscow.

Goncharova, E.P. 2010. Biodegradable electret films on the base of polythene for packing goods of light industry. Ph.D. Thesis, Belarus State Technological University, Minsk.

Goncharova, E.P., O.A. Ermolovich., Pinchuk L.S. and V.E. Sytsko. 2006. Modern trends in biodegradable polymeric materials creation. Material Science 9: 37–43.

Goncharova, E.P., V.E. Sytsko, L.S. Pinchuk. 2009. Repellent biodegradable film for packing textile and fur products. Materials. Technologies. Tools 2: 51–54.

Gould, P. 2004. Nanoparticles probe biosystems. Materials Today 7/2: 36–44.

Grande, J.A. 1997. Biocides invade large consumer good market. Modern Plastics Int. 27/9: 61.

Grigor'ev Yu.G. (ed.). 2002. *Electromagnetic Fields and Human Health*. Russian University of Nations Friendship, Moscow.

Grigor'yants, A.G., I.N. Shiganov and A.I. Misyurov. 2006. *Technological Processes of Laser Processing: School-book*. N.E. Bauman Moscow State Technical University, Moscow.

Gubkin, A.N. 1978. *Electrets*. Science, Moscow.

Gunther, G.J. and G. Beck (eds.). 1983. *Metal Glasses*. Mir, Moscow.

Guzman, I.Ya. 1995. Heat-insulating materials. pp. 525–526. *In*: N.S. Zefirov (ed.). *The Chemical Encyclopedia, Vol. 4*. Big Russian Encyclopedia, Moscow.

Gyunter, V.E., G.C. Dambaev, P.G. Sysolyatin, R.V. Ziganshin and M.Z. Mirgazaev. 1998. *Medical Materials and Implants with Memory of the Form*. Tomsk State University, Tomsk.

Hara, K., K. Aoki, Yu. Usui, M. Shimizu, N. Narita, N. Ogihara, K. Nakamura, N. Ishigaki, K. Sano, H. Haniu, H. Kato, N. Nishimura, Y.A. Kim and S. Taruta. 2011. Evaluation of CNT toxicity by comparison to tattoo ink. Materials Today 14/9: 434–440.

Hardy, J.W. 1977. Adaptive optics. J. Opt. Soc. Amer. 67/3: 21–33.

Heinicke, G. 1984. *Tribochemistry*. Akademie-Verlag, Berlin.

Hiroyasu, Yu. and N. Masanobu. 2006. Superconducting Cable. Russia Patent # 2,340,969.

Hurris, J.M. (ed.). 1992. *Polyethylenglicol Chemistry, Biotechnical Biomechanical Applications*. Plenum Press, New York.

Ilizarov, G.A. 1952. Process of bones joining at fracture and the device for its realization. USSR Inventor's certificate # 98,471.

Illarionov, V.E. 2001. *Conceptual Bases of Physiotherapy in Rehabilitology: A New Paradigm of Physiotherapy*. Medicine, Moscow.

Isakovich, G.A. and G.L. Osipov. 1970. Acoustic materials. pp. 367–368. *In*: A.M. Prokhorov (ed.). *Big Soviet Encyclopedia, Vol. 1*. Soviet Encyclopedia, Moscow.

Ivanov, B.A. 1999. Refrigerating processes. pp. 593–605. *In*: N.S. Zefirov (ed.). *The Chemical Encyclopedia, Vol. 5*. Big Russian Encyclopedia, Moscow.

Jacob, A. 2002. Self-healing FRP up for license. Reinforced Plastics 46/9: 18.

Jahanmir, S. (ed.) 1984. *Friction and Wear of Ceramics*. Marcel Dekker, New York.

Jang, J. 2006. Displays develop a new flexibility. Materials Today 9/4: 46–52.

Karlov, A.V. and V.P. Shakhov. 2001. *Systems of External Fixing and Adjusting Gears of Optimal Biomechanics*. STT, Tomsk.

Kazachkovskii, O.D. 1998. Fuel element. p. 53. *In*: A.M. Prokhorov (ed.). *The Physical Encyclopedia, Vol. 5*. Big Russian Encyclopedia, Moscow.

Klassen, V.I. 1982. *Magnetic Treatment of Water Systems*. Chemistry, Moscow.

Kogan, L.M., S.M. Kovykin, V.S. Rodkin and Yu.P. Andreev. 1990. New light-emitting diodes. Electronic Industry 9: 142.

Konek, D.A., K.V. Voitsekhovski, Yu.M. Pleskachevskii and S.V. Shilko. 2004. Materials with negative Poisson's ratio (review). Mechanics of Composite Materials and Designs 10/1: 35–69.

Kopytkov, V.V. 2007. Polymeric compounds for processing root systems of pine seedlings. Ph.D. Thesis, Belarus State Technological University, Minsk.

Korotkii, M.V. 2005. Modifying in electric and magnetic fields of fibrous polymeric biomass carriers for aerobic biofilters. Ph.D. Thesis, Belarus State Technological University, Minsk.

Korotkii, M.V., A.V. Makarevich and L.S. Pinchuk. 2004. Polymeric carriers of microorganisms – sources of biogenic elements. Material Science 11: 52-56.

Kostyukov, N.S., N.P. Antonova, M.I. Zilberman, N.V. Aseev. 1979. *Radiation Electro Material Science*. Atomizdat, Moscow.

Kozlov, L.V. Immobilized enzymes. 1990. pp. 421–423. *In*: I.L. Knunyants (ed.). *Chemical Encyclopedia, Vol. 2*. Soviet Encyclopedia, Moscow.

Kravtsov, A.G., V.A. Goldade, and S.V. Zotov. 2003. *Polymeric Electret Filtering Materials for Respiratory Organs Protection*. Metal-Polymer Research Institute of Belarus NAS, Gomel.

Kuryshev, K. 2001. Polymeric safety devices PolySwitch – reliable way to protect electric circuits from damages. Components and Technologies 2: 41–45.

Kuzmin, M.G. 1999. Photochemical reactions. pp. 350–356. *In*: N.S. Zefirov (ed.). *The Chemical Encyclopedia, Vol. 5*. Big Russian Encyclopedia, Moscow.

Kuznetsov, I.I. 1980. Way of increase of man's serviceability and the device for its realization. USSR Inventor's certificate # 791,387

Lazarev, G.E. 1987. The main regularities of corrosion-resistant steels and alloys wear at friction in electrolytes. J. of Friction and Wear 8/2: 223-230.

Lazarev, G.E. and G.A. Preis. 1979. Friction and wear in corrosive media. pp. 308–324. *In*: I.V. Kragelskii and V.V. Alisin (eds.). *Friction, Wear and Lubrication: Reference Book, V. 2*. Mechanical Engineering, Moscow.

Lazarev, V.B., V.G. Krasov and I.S. Shaplygin. 1978. *Electroconductivity of Oxide Systems and Film Structures*. Science, Moscow.

Leaversuch, R. 2001. Polyolefins find niche in IV bags. Modern Plastics Int. 31/5: 96–97.

Litvinov, V.N., N.M. Mikhin and N.K. Myshkin. 1979. *Physical and Chemical Mechanics of Selective Transfer at Friction*. Science, Moscow.

Liu, Zh., J.T. Robinson, S.M. Tabakman, K. Yang and H.J. Dai. 2011. Carbon materials for drug delivery and cancer therapy. Materials Today 14/7-8: 316–323.

Losk, F. 1999. *Packing and Ecology*. Moscow State University of Foodstuffs, Moscow.

Lyapko N.G. 2007. The device for reflexotherapies (variants). Russia Patent # 2,304,957

Ma, P.X. and J. Elisseeff (eds.) 2005. *Scaffolding in Tissue Engineering*. Tailor & Francis CRC Press, London.

Makarevich, A.V. 2000. Physical, chemical and technological principles of creation of active film and fibrous materials on the base of thermoplastic polymers. Dr. Sci. Thesis, Belarus State Technological University, Minsk.

Makarevich, A.V., L.S. Pinchuk and V.A. Goldade. 2003. Electric Fields and Electroactive Materials in Biotechnology and Medicine. Metal-Polymer Research Institute of Belarus NAS, Gomel.

Makarov, V.A. 1999. Electrochemical protection. pp. 907–910. *In*: N.S. Zefirov (ed.). *The Chemical Encyclopedia, Vol. 5*. Big Russian Encyclopedia, Moscow.

Mandelbaum, B., J. Browne, F. Fu, L. Micheli, J.B. Mosely, Jr.C. Erggelet, T. Minas, and L. Peterson. 1998. Articular cartilage lesions of the knee. Am. J. Sport Med. 6: 853–860.

Marsh, G. 2002. Composites strengthen aerospace hold. Reinforced Plastics 46/7-8: 40–43.

Martin, S. 1999. Food-approved. Modern Plastics Int. 29/11: 89–90.

Matveevskii, R.M. 1978. Temperature stability of boundary lubricant layers and oiling film coatings. pp. 296–307. *In*: I.V. Kragelskii and V.V. Alisin (eds.). *Friction, Wear, Lubrication, Vol. 1*. Mechanical Engineering, Moscow.

Merkulov, V.I. and S.K. Golushko. 2009. Capacitive current generator. Russia Patent # 2,346,380.

Minich, A.S. and V.S. Raida. 1998. Laboratory method of service life testing of phosphor in photocorrecting films. Plastics 5: 34.

Ministr, Z. and J. Priester. 1970. New dry-running bearing material. pp. 303–308. *In*: *Perspectives in Powder Metallurgy. Friction and Antifriction Materials*. Plenum Press, New York.

Mironov, V.L. 2004. *Foundation of Scanning Probe Microscopy*. Teckhnosphere, Moscow.

Mironov, V.S. and Yu.M. Pleskachevskii. 1999. *Electrophysical Activation of Polymer Materials*. Metal-Polymer Research Institute of NAS of Belarus, Gomel.

Mogg, R. 2001. Processing aids and lubricants. Modern Plastics Int. 31/8: 59–60.

Morgunov, M.S., V.P. Homutov and J.M. Sokolova. 1994. Application of electrets in traumatology and orthopedy. Proc. 8th Int. Symp. on Electrets, Paris: 863–868.

Moskvitin, V.V. 1981. *Cyclic Loading of Constructions Elements*. Science, Moscow.

Nagasaki, Yu. and K. Kataoka. 1996. An intelligent polymer brush. Trends in Polym. Sci 4/2: 59–64.

Netravali, A.N. and Sh. Chabba. 2003. Composites get greener. Materials Today 6/4: 22–29.

Neverov, A.S., Rodchenko D.A. and M.I. Tsyrlin. 2007. *Corrosion and Protection of Materials*. Higher School, Minsk.

Nicholson, J. 1998. Current trends in biomaterials. Materials Today 1/2: 6–8.

Nikulin, M.A. 1990. Trophic disease in the view of electromagnetic interactions. pp. 116–121. *In: Actual Problems of Application of Magnetic and Electromagnetic Fields in Medicine*. Science, Leningrad.

Nobel laureates. 2007. *The Discovery of Giant Magnetoresistance. Scientific background on the Nobel Prize in physics 2007*. The Royal Swedish Academy of Sciences, Oslo.

Nowick, A.S. and B.S. Berry. 1972. *Anelastic Relaxation in Crystalline Solids*. Academic Press, New York

Noy, A., A.B. Artyukhin and N. Misra. 2009. Bionanoelectronics with 1D materials. Materials Today 12/9: 22–31.

Oraevskii, O.N. 1990. The laser. pp. 546–552. *In*: A.M. Prokhorov (ed.). *The Physical Encyclopedia, Vol. 2*. Soviet Encyclopedia, Moscow.

Orlov, D.V. 1977. The theory, research and development of electromechanical sealing systems with liquid-metal and magneto-liquid working substance. Dr. Sci. Thesis, Moscow Power Institute, Moscow.

Papkov, S.P. 1981. *Phase Equilibrium in Polymer-Solvent System*. Chemistry, Moscow.

Pasic, R.P. and R.L. Levine. 2007. *A Practical Manual of Laparoscopy and Minimally Invasive Gynecology*. Informa Healthcare, London.

Pasynkov, V.V., K.L. Chirkin and A.D. Shinkov. 1987. *Semiconductor Devices*. Higher School, Moscow.

Patterson, J., M.M. Martino and J.A. Hobbell. 2011. Biomimetic materials tissue engineering. Materials Today 13/1-2: 14–22.

Penkin, N.S. 1977. *Gummed Details of Machines*. Mechanical Engineering, Moscow.

Perepelkin, K.E. 1998. Enzyme-containing fibers. pp. 157–158. *In*: N.S. Zefirov (ed.). *Chemical Encyclopedia, Vol.5*. Big Russian Encyclopedia, Moscow.

Perez-Castillejos, R. 2011. Replication on the 3D architecture of tissues. Materials Today 13/1-2: 32–41.

Pertsov, A.V. 1995. Thin films. pp. 1206-1207. *In*: N.S. Zefirov (ed.). *The Chemical Encyclopedia*, Vol. 4. Big Russian Encyclopedia, Moscow.

Petryanov, I.V., V.I. Kozlov, P.I. Basmanov and B.I. Ogorodnikov. 1963. *Fibrous Filtering Materials* FP. Knowledge, Moscow.

Petty, W (ed.). 1991. *Total Joint Replacement*. W.B. Saunders Co, Philadelphia.

Pimenov, A.F. (ed.). 1990. *Pressure Treatment of Metal Materials*. Science, Moscow.

Pinchuk, L.S. 1992. *Hermetology*. Science and Engineering, Minsk.

Pinchuk, L.S. and A.S. Neverov. 1993. *Polymeric Films Containing Corrosion Inhibitors*. Chemistry, Moscow.

Pinchuk, L.S. and V.A. Goldade. 2014. *Crazing in Technology of Polyester Fibers.* Belarus Science, Minsk.

Pinchuk, L.S., A.G. Kravtsov, Yu.I. Voronezhtsev and Yu.V. Gromyko. 1998. On the charge state of melt-blown polymer material. Int. J. Polymer Processing 13/1: 67–70.

Pinchuk, L.S., A.V. Makarevich and V.A. Goldade. 2001a. Active polymeric packing films. Technology of Processing and Packing 4: 30–33.

Pinchuk, L.S., and A.S. Neverov. 1995. *Sealing Polymer Materials.* Mechanical Engineering, Moscow.

Pinchuk, L.S., E.A. Tsvetkova and E.M. Markov. 1992. Magnetotherapeutic device. USSR Inventor's certificate # 1,776,402.

Pinchuk, L.S., E.A. Tsvetkova and Zh.V. Kadolich. 2001b. Influence of electromagnetic fields on friction in joint implants. J. Friction and Wear 22/5: 550–554

Pinchuk, L.S., I.Yu. Ukhartseva, E.I. Parkalova and O.I. Pashnin. 2001c. Wrapping film for cheese. Packing 4: 19–21.

Pinchuk, L.S., M.V. Korotkii and E.P. Goncharova. 2009. The bioelectric mechanism of microorganisms immobilization on polymeric electret films. Reports of Belarus NAS 53/2: 107–110.

Pinchuk, L.S., O.I. Pashnin, V.A. Goldade, A.A. Lyubin, A.V. Makarevich, I.Yu. Ukhartseva, R.S. Napreev and E.I. Parkalova. 2005. Multilayered film for cheeses packing and maturing. Russia Patent # 2,250,831

Pinchuk, L.S., V.A. Goldade and E.A. Tsvetkova. 1995. The device for segmentary reflexotherapy and its variants. Belarus Patent # 754.

Pinchuk, L.S., V.A. Goldade and S.Ya. Liberman. 1991. Polymeric anticorrosion film. U.S. Patent # 5,028,479.

Pinchuk, L.S., V.A. Goldade, A.V. Makarevich and V.N. Kestelman. 2002. *Melt Blowing: Equipment, Technology and Polymer Fibrous Materials.* Springer, Berlin.

Pinchuk, L.S., V.I. Nikolaev, E.A. Tsvetkova and V.A. Goldade. 2006. *Tribology and Biophysics of Artificial Joints.* Elsevier, London.

Pinchuk, L.S., Yu.M. Chernyakova and E.A. Tsvetkova 2011. Way of processing of polymeric head of unipolar prosthesis of a joint. Belarus Patent 14,349.

Pkhakadze, G.A. 1990. *Biodegradable Polymers.* Scientific Thought, Kiev.

Pleskachevskii, Yu.M., V.V. Smirnov and V.M. Makarenko. 1991. *Introduction to Radiation Material Science of Polymer Composites.* Science and engineering, Minsk.

Poller, Z. 1982. *Chemistry on the Way to the Third Millennium.* Mir, Moscow.

Potekha, V.L. 2000. *Tribodilatometry.* Gomel State Technical University, Gomel.

Pratt, J.S. 1993. Bearing alloys for engines of internal combustion. pp. 312–330. *In*: *Tribology: Researches and Applications. Experience of the USA and CIS Countries.* Mechanical Engineering, Moscow.; Allerton Press, New York.

Preis, G.A. and A.G. Dzub. 1980. Electrochemical phenomena at friction of metals. J. Friction and Wear 1/2: 215–235.

Prince, K. 1998. The Austrian plastics industries. Modern Plastics Int. 28/6: 99.

Rainwater, D., A. Kerkhoff, K. Melin, J.C. Soric, G. Moreno and A. Alù. 2012. Experimental verification of three-dimensional plasmonic cloaking in freespace. New J. of Physics 14, Jan.: 77–82

Rakhubovskii, V.A. 2004. Cryotron devices. Matters of Nuclear Science and Engineering 6: 104–106.

Reece, P.L. 2007. *Progress in Smart Materials and Structures.* Nova Science Publishers, New York.

Rehbinder P.A. 1978. Surface Phenomena in Disperse Systems. Science, Moscow.

Reitlinger, S.A. 1974. *Permeability of Polymeric Materials*. Chemistry, Moscow.

Rozenberg, V.M. 1994. Materials Creep. pp. 10–13. *In*: A.M. Prokhorov (ed.). *The Physical Encyclopedia, Vol. 4*. Big Russian Encyclopedia, Moscow.

Sailor, M.J. and J.R. Link. 2005. Smart dust: nanostructured devices in a grain of sand. Chemical Communications 11: 1375–1380

Savich, V.V., M.G. Kisilev and A.I. Voronovich. 2003. *Modern Materials of Surgical Implants and Tools*. Tekhnoprint, Minsk.

Scarborough, N. and R.A. Griend. 1991. Allografts. pp. 43-50. *In*: W. Petti (ed.). *Total Joint Replacement*. Saunders Co., Philadelphia.

Selkin, V.P. 2004. Use of dosimeters the based on cross-linked polyethylene in industrial radiation processes. Materials. Technologies. Tools 9/3: 110–112.

Semlitsch, M. 1992. 25 years Sulzer development of implant materials for total hip prostheses. Sulzer Medical Journal 2: 23–27.

Sessler, G.M. and J.E. West. 1987. Applications of electrets. p.p. 347–381. *In*: G.M. Sessler (ed.). *Electrets*. Springer-Verlag, Berlin.

Shalaev, V.M. and A.K. Sarychev. 2007. *Electrodynamics of Matamaterials*. World Sci. Publ. Co., New York.

Sharapov, V.M., M.P. Musienko and E.V. Sharapova. 2006. *Piezoelectric Sensors*. Technosphere, Moscow.

Shatt, V. (ed.). 1983. *Powder Metallurgy. Sintered and Composite Materials*. Metallurgy, Moscow.

Shishatskaya, E.I., T.G. Volova, S.A. Gordeev and A.P. Puzyr. 2002. Biodegradation of sutural fibers on a base of polyoxyalkanoates in biological media. Perspective Materials 2: 56–62.

Shlyapnikov, Yu.A. 1990. Destruction of polymers. pp. 38–41. *In*: I.L. Knunyanc (ed.). *The Chemical Encyclopedia, Vol. 2*. Soviet Encyclopedia, Moscow.

Shulman, Z.P., Yu.F. Deinega, R.G. Gorodkin and A.D. Macepuro. 1972. *Electrorheological Effect*. Science and Engineering, Minsk.

Shvarts, S., J. Shaiers and F. Spenser (eds). 1999. *The Directory on Surgery*. Piter Press, St.-Petersburg.

Sigidin, A.Ya. 1988. *Medicinal Therapy of Inflammatory Process*. Medicine, Moscow.

Silin, A.A. 1976. *Friction and Its Role in Engineering Development*. Science, Moscow.

Simonova, L.N. 1998. Indicators. pp. 446–452. *In*: N.S. Zefirov (ed.). *Chemical Encyclopedia, Vol. 5*. Big Russian Encyclopedia, Moscow.

Sinitsyn, G.V. 1987. Completely optical elements of discrete logic on the base of bistable thin-film interferometers. Quantum Electronics 14/3: 529–538.

Smyslov, I. 2002. New opportunities of high technology: Posistor heaters. Electronics: Science. Technology. Business 4: 18–20

Sokolskii, Yu.M. 1990. *Magnetic Treated Water: the Truth and Fiction*. Chemistry, Leningrad.

Solntsev, Yu.P., E.I. Pryakhin and F. Voitkun. 1999. *Material Science*. The Moscow Institute of Steel and Alloys, Moscow.

Sonin, A.S. 1983. *Introduction in Liquid Crystals Physics*. Science, Moscow.

State standard of Russia # 17527. 1986. *Packing. Terms and definitions*. State Standard Publ., Moscow

Stewart, R. 2002. US Navy enlists Swedish 'stealth' technology. Reinforced Plastics 46/12: 8

Struk, V.A., L.S. Pinchuk, N.K. Myshkin, V.A. Goldade and P.A.Vityaz. 2010. *Material Science in Mechanical Engineering and Industrial Technologies*. Intellect, Moscow.

Suzdalev, I.P. 2009. Multipurpose nanomaterials. *Russian Chemical Reviews* 78/3: 266–301

Sychev B.S. 1994. Radiative stability of materials. pp. 202–203. *In*: A.M. Prokhorov (ed.). *The Physical Encyclopedia, Vol. 4*. Big Russian Encyclopedia, Moscow.

Tada, Ya. 1992. Experimental characteristics of electret generator using polymer film electrets. Jap. J. Appl. Phys. 31: 846–851.

Teplyakov, A. 2006. Defend yourself; sirs! Science and Innovations 9: 14–17.

Tishin, A.M. 2007. Process of carrying out of magnetic therapy of malignant growth. Russia Patent # 2,295,933.

Tishin, A.M. and Yu.I. Spichkin. 2005. The magnetic thermal machine. Russia. Patent # 2,252,375.

Todd, A., J. Busby, M. Meyer and D. Petti. 2010. Materials challenges for nuclear systems. Materials Today 13/12: 14–23.

Tomer, B. 1999. Plastics cork finds its niche among world-class vintners. Modern Plastics Int. 29/11: 22.

Tretyakov, S. 2003. *Analitical Modeling in Applied Electromagnetics*. Artech House, Norwood.

Tuchkevich, V.M. and I.V. Grekhov. 1988. *New Principles of Large Capacities Switching by Semiconductor Devices*. Energoatomizdat, Leningrad.

Ulijn, R.V., N. Bibi, V. Jayawarna, P.D. Thornton, S.J. Todd, R.J. Mart, A.M. Smith and J.E. Gough. 2007. Bioresponsive hydrogels. Materials Today 10/4: 40–48.

Uskov, M.K. and E.F. Bogdanov (eds.). 1995. *Mechanical Engineering: the Terminological Dictionary*. Mechanical Engineering, Moscow.

Van Geuns, J.R. 1968. Method and apparatus for cold producing. U.S. Patent # 3,413,814.

Varfolomeev, S.D. and K.G. Gurevich. 1999. *Biokinetics: Practical Curriculum*. Fair-Press, Moscow.

Varivodov, V.N. 2008. New technologies for Russian power companies. Energy Saving 4: 32–39.

Vasnev, V.A. 1997. Biodegradable polymers. High-molecular Compounds B12: 2073–2086.

Veselago, V.G. 1967. Electrodynamics of substances with homogeneous negative values ε and μ. Physics-Uspekhi (Advances in Physical Sciences) 92: 517–523.

Vinchester, A. 1967. *Foundation of Modern Biology*. Mir, Moscow.

Volkenshtein A.A. and E.V. Kuvaldin. 1975. *Photo-Electric Pulse Photometry*. Energoizdat, Leningrad.

Voronezhtsev Yu.I., V.A. Goldade, L.S. Pinchuk and V.V. Snezhkov. 1990. *Electric and Magnetic Fields in Technology of Polymeric Composites*. Science and Engineering, Minsk.

Vorontsov, A.F., E.S. Demin and S.B. Demin. 2010. Magnetostrictive two-coordinate tiltmeter. Russia Patent # 2,389,975.

Vysotskii, B.F. and V.V. Dmitriev (eds.). 1985. Integrated Piezoelectric Devices of Signals Filtration and Processing. Radio and Communication, Moscow.

Wareing, P.F. and I.D.J. Phillips. 1981. *Growth and Differentiation in Plants*. Pergamon Press, Oxford.

Williams, D.F. (ed.). 1987. *Definitions in Biomaterials*. Elsevier, Amsterdam.

Wolf, S.A., D.D. Awschalom, R.A. Buhrman, J.M. Daughton, S. von Molnar, M.L. Chtchelkanova and D.M. Treger. 2001. Spintronics: a spin-based electronics vision for the future. Science 294/5546: 1488–1495

Yamamoto, Yo., T. Fukushima and Yu. Suna. 2006. Photoconductive coaxial nanotubes of molecularly connected electron donor and acceptor layers. Science 314/5806: 1761–1764.

Yanagida, H. (ed.). 1986. *Thin Technical Ceramics*. Metallurgy, Moscow.

Zeeshan, M.A., K. Shou, K.M. Sivaraman, T. Wuhrmann, S. Pané, E. Pellicer and B.J. Nelson. 2011. Nanorobotic drug delivery. Materials Today 14/1-2: 54.

Zhdanov, B.V. 1992. Light shutter. p. 453. *In*: A.M. Prokhorov (ed.). *The Physical Encyclopedia, Vol. 5*. Big Russian Encyclopedia, Moscow.

chapter three

Mechanics of smart materials

3.1 Object and subject of smart materials mechanics

The history of science and technology bear evidence that a preliminary analysis of the problem using the mechanics and mathematics models facilitates the discovery of new effects and regularities, the birth of technical solutions and the promotion of innovations.

The methods of deformable solid mechanics and those of the related disciplines (physical mesomechanics, biomechanics, mechanochemistry) and the CAD technology (computer-aided design, virtual experiments) became instruments for the SM developers. It is felt that high molecular weight species (macromolecular compounds) and first and foremost – polymers, oligomers, and composites based thereon – allow much room for the creation/development of new materials having signs of artificial intelligence (Galaev1995, Burukhin and Babkin 1995, Pleskachevsky et al 1999). These high molecular weight species among all the artificial materials media are mostly closed to Nature. Artificial polymers and oligomers have high sensitivity (fast response) to the physical fields that proves their suitability to perform functions of sensors and actuators.

The hierarchy of scale levels and mobility of uniformly 'built' biomaterials' structural elements and synthetic polymers is expressed in the spectrum of mechanochemical, magnetic and electrophysical effects. It is possible on the basis of these effects to implement different operations such as, generation, commutation, amplification and detection of signals, information storing and physical fields' transformation as derivative elements of SM sensory, processor and actuator functions. No doubt that high molecular weight compounds are the only material carriers of adaptive responses. Metal alloys, ferroelectric ceramics and other materials with special properties may be their rivals in this respect. Biological structures with the self-organizing ability generated in the course of their evolution are SM prototypes (Paturi 1974). All these marked features in aggregate with the obligatory presence of feedback in the smart materials define the subject of SM mechanics research, including:

- description of material deformation and fracture processes at the mesolevel ("meso" – medium, intermediate; "mesostructure"

– intermediate between the phase and the atom or molecule, such as a crystal or supramolecular structures);
- determination of bifurcation points (bifurcation – the material acquisition of new quality at small change in its structure), cooperative and self-organization effects on the 'property – external influence' dependence;
- formulation of the adaptability criteria and objectives concerning movable (free) external and interphase boundaries;
- calculation of critical (threshold) parameters values at which the material adaptation to external influence and FBS 'switching on' take place.

The authors do not claim full coverage of the bibliography emerged in recent years on the subject of SM mechanics research (only on self-healing defects each year appears a few dozen references).

3.2   Structural and functional analysis of SM in terms of mechanics

3.2.1 Role of polymer composites as the precursors of SM

In the view of mechanics, structural and functional analysis of composites is based on the deformation and strength properties experimental data, as well as phenomenological simulation and multi-level discrete models of composites, taking into account non-linear dependencies of the physical and mechanical properties on external conditions and the interaction of the components on the interface. System of deformable solid mechanics equations characterizing the properties of polymer composites is apparently the most complicated, though heterogeneity and anisotropy of properties is inherent in composites of any nature (Table 3.1).

The macromolecules specific configuration resulting in the ability of polymers and elastomers for the large elastic deformations that, along with their inherent creep and orientation reinforcement, the presence of special temperature points (glass transition, crystallization, yield, brittleness, thermal destruction), significantly complicates analysis and modeling for them. Interpretation of the listed phase transitions and conformational macromolecules changes requires the use of quantum mechanics conceptions.

The constructional material operating capacity is determined by its set of deformation and strength characteristics, the most important of which are the specific modulus of elasticity E/ρ and specific tensile strength σ/ρ (Fig. 3.1). According to data and prediction available at the end of the last century (Fudzii and Dzako 1982) it is evident that specific strength σ/ρ of

Table 3.1. Comparison of physical and mechanical properties for composites on different matrices

Characteristic	Matrix material		
	metal	ceramics	polymer
Heterogeneity	*	*	***
Anisotropy	*	*	***
Creep, relaxation	*	0	***
Rate sensitivity of mechanical characteristic	*	*	***
Orientational reinforcement	**	*	***
Large elastic strains	0	0	***
Incompressibility	0	0	***
Temperature points of properties sharp change	*	*	***

The degree of properties manifestation: *** – high, ** – average * – moderate, 0 – absent.

steel, aluminum, titanium and other metals has not undergone radical changes.

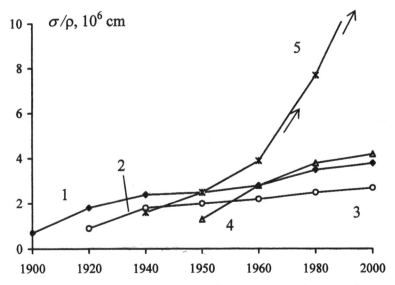

Fig. 3.1 Increasing of specific tensile strength of materials in last period; extrapolation of data (Fudzii and Dzako 1982): *1* – steel, *2* – aluminum, *3* – titanium, *4* – boron and carbon plastics

In accordance with the theory of dispersion strengthening, yield limit of the composite material filled with solid particles increases in inverse proportion to the distance between them. All-time high characteristics of the best desired metal structural material – steel having an average particle size of cementite about 100 nm have been achieved by a multistage rolling and subsequent heat treatment: annealed tensile steel ultimate strength was about 900 MPa (plasticity up to 15%); ultimate compression strength of hardened steel – 4500 MPa (10% plasticity).

Reinforcement of polymer matrix with glass, boron and carbon fibers provides specific characteristics of composite material unattainable for metals and alloys. Moreover, the strength potentialities of such materials have not yet been exhausted. The nanotechnology progress confirms this thesis: the modification of filled plastics with carbon nanotubes, having high strength and low coefficient of thermal expansion, provides super high strength of metals and polymers.

The next stage of materials science development will deal with the intelligent (smart) materials ("mechanocomposites" in terms of this Part) that don't obligatory have *a priory* extreme high physical and mechanical characteristics, but can adapt themselves to the external influence. Potentially high operating capacity of adaptive materials is reached with their flexible structure: mobility of interphase boundaries, commutating of local cohesive and adhesive bonds, controlling of the density and elasticity modulus. The FBS functioning provides a self-test function, the healing of defects, self-strengthening in the extreme conditions, etc. This will minimize the probability of machinery failure due to damage of materials, improve the reliability of structures and facilitate their disposal.

The Fig. 3.2 and Table 3.2 illustrate the typical objects (natural and artificial inhomogeneous and functional materials, problems and methods of composites mechanics with reference to SM.

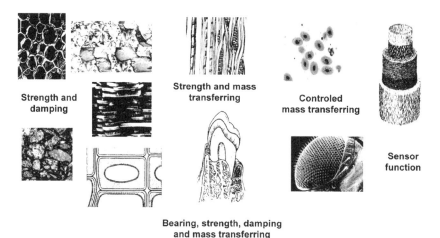

Strength and damping

Strength and mass transferring

Controled mass transferring

Sensor function

Bearing, strength, damping and mass transferring

Fig. 3.2 Natural and artificial composites as inhomogeneous and functional materials

Table 3.2. Problems and methodology of composites mechanics with reference to smart materials

Problem	Methods
Description of crystals, eutectics, supra-molecular systems, 'filler-interface-matrix' structural units	Finite and boundary elements discretization, fractal theory, hierarchic adaptive models
Interpretation of aggregate and phase states	Quantum mechanical models, cell automata, self-organization principles
Description of degradation and creep of materials	Theory of relaxation in non-equilibrium structures
Strength and deformability prediction of SM based issues	Multi-level models of stress-strain state and failure kinetics taking into account the feedback
Determination of full set of elasticity constants, yield and strength limits for SM mesofragments	Tomography, probe analysis on micro-scale level, procedures of models identification
Development of SM technology and design of issues having feed back	Object-oriented data bases and expert systems of materials reliability, local and global information networks
Combination of mathematical modeling and physical experiment	Numerical methods and discretization of experimental data

The experiment and theory integration in the area under consideration as it is reflected in Table 3.2 means the following. Until recently, canonical and fairly simple models of deformation (linear elasticity, rigid-plastic or viscoelastic flow, etc.) seemed to be universal for most of the materials. For the rheologically complex SM, however, types and limits of state equations applicability require substantial revision. The SM deformation behavior is extremely nonlinear that limits using of the traditional elastic constants, viscoelastic and viscoplastic parameters, velocity sensitivity etc. acceptable in the case of conventional, relatively stable materials for SM description. It is advisable to carry out the SM strength analysis in real time on the deformable volume with moving interphase boundaries on space discretization basis. This requires application of the high-precision automated testing-measuring equipment and computer programs, presumes matrix and operator representation of the initial data. It is not out of place here to give an analogy of smart material with modern mechanical testing machine having in addition to the loading and force measuring devices a number of sensors and a computer system for stabilization of forces, movements and heat flow having an effect on the sample.

3.2.2   Macromechanical modeling of SM

The methods currently used in the mechanics of composites are based
on the replacement of the composite with the equivalent homogeneous
medium having known characteristics (macromechanical approach)
(Sendeckyj 1974, Shermergor 1977, Christensen 1979). It is supposed that
volume of the material under study is considerably larger than the characteristic phase inhomogeneity (particles, fibers, pores or other inclusions).
Selecting of a continuum model (describing continuous medium) and its
verification (proof of authenticity) is performed on the experimental data
basis. Effective modules, i.e. the coefficients relating volume-averaged
components of the stress and strain tensors, are determined on the basis
of standard mechanical tests, trying to take into account the phase inclusions nature, their elastic characteristics, shape and location. After that, a
multi-component medium can be described by the generalized Hooke's
law

$$\sigma = C\varepsilon, \tag{3.1}$$

where the elasticity tensor C is piecewise constant with respect to a separate phase. Knowing the coefficients of the composite tensor C as a whole,
we can find the elastic moduli used in engineering practice.

Effective tensors of the elastic and compliance moduli are represented
in the form of a series corresponding to concentration of material components under low volumetric content of one of it in the material. The
approach disadvantage is that the elasticity constants of the multicomponent material are taken equal to the elasticity constants of the matrix
not only near the phase inclusion, but at large distances from it as well.
Obviously, elastic properties of the material correspond to the effective
elasticity moduli of the composite as a whole at a large distance from the
phase grain.

Self-matching method is that a transitional interphase layer is singled
out in the material. The thickness of the layer is about a few medium
size phase inclusions. The continuous medium elastic characteristics are
described by the effective elasticity constants of the whole material outside the layer. Self- matching method approximation includes change of
the grain's real form to the canonical one and the transition of the interphase layer to a zero thickness.

Correlation approximation of random functions theory leads to good
results if there is a small difference in the elasticity moduli of the material components. It takes place in the polymer-polymer composites, when
the calculations can use two small parameters – the relative difference
between the moduli of elasticity and compliance. The relevant solution
is obtained in the form of an equation expansion for these parameters.
The random functions theory allows representing the effective tensors of

elasticity and compliance moduli as the sum of the mean value and correlation corrections taking into account the multiple interactions between the zones of heterogeneity.

Variational method for calculation of the effective elasticity moduli of a composite material is based on the principle of minimum complementary energy expended for the deformation and it provides a more accurate assessment of the elastic constants. In general, solutions are obtained for the combined thermoelasticity and thermoviscoelasticity problems, kinetic and statistical models of inelastic deformation and fracture of composite materials in the framework of the multi-component medium macroscopic examination. The development of hierarchical models providing clarification of the viscoelastic state equations based on the selected criterion of the quality of the material may be attributed to the latest developments in this area (Basistov and Yanovskii 1996, Oden et al 1999).

The modern level of the composite material mechanics allows analyze stresses, displacements and strains in the area of arbitrary shape inclusions arising due to simultaneous influence of power, thermal, electromagnetic and radiation fields. It is possible to 'play' different scenarios of deformation and fracture, 'healing' of defects, phase transitions and structural transformations with the help of simulation modeling that is essential in describing of the adaptive and smart composites. With this aim in view, beside the macromechanical analysis the other structural level detailing methods are developing in recent years. These are largely interdisciplinary approaches for the mechanical properties of materials unfolding.

3.2.3   *Physical mesomechanics methods for SM modeling*

Methods of physical mesomechanics (Panin 1995, Panin 2010) actualize the R. Descartes famous aphorism "Divide any problem into as many parts as it is possible" and his paradigm of 'motion vector analysis'. These methods allow one to describe the formation of the mechanical properties of the material as mutually coordinated behavior of its structural units – grains, fibers, crystallites and even supramolecular structures. 'Meso' means intermediate scale level allocated by the researcher; this level's absolute sizes may differ in several orders of magnitude. Thus, the plastic deformation of polycrystals process was interpreted in the terms of bands shift (Panin 1995). At the same time, in modeling of seismic events, their mesoscale correspond many kilometers cracks in ice cover.

The hierarchical self-organization of interrelated non-uniformly scaled intermediate levels of deformation determines the properties of solids in the various fields of external influences, as of now, it seems to be the most realistic approach. As a part of the physical understanding the deformation behavior of material (no matter – is it passive, active or intelligent), mesomechanics is associated with representations of strength physics, mechanochemistry (Casale and Porter 1978) and macromolecular chemistry (Berlin et al 1992).

In the nearest future, this science implemented as a computer-aided technology for new materials development will combine capabilities of mechanics with physical and chemical principles of smart composite designing.

The idea of modeling disordered solid bodies with homogeneous structures, for instance with finite elements, with mobile cellular automata (Toffoli and Margolus 1987), and others is used in the mesomechanics. A repeating structural unit (a periodicity cell) is thus the key concept of mesomechanics for materials. Identification of local deformations and configuration of interphase boundaries is required for the tracing of a physically more meaningful mesomechanical model alternative to phenomenological description of the composite construction. This problem can be solved with the help of the models based on the structural units, these sells examples for various heterogeneous (porous, granular, dispersion-filled) materials are shown in Fig. 3.3.

The mesomechanical approach advantage is the possibility of consecutive detailing of 'inexhaustible' real structure by allocating new scale levels. Thereby the concepts of 'material' and 'construction' are generalized, as far as a heterogeneous material is interpreted as a system of elements that performs specific functions.

The number of levels chosen to describe the inhomogeneous materials should correspond to the degree of detailing, optimal in the view of the model adequacy to the process under consideration and to the calculation means available (Fig 3.4).

Very intricate problems arise in modeling of the active and adaptive materials having artificial and biological origin, for example, such as muscle tissue with a multilayered structure and capable of a force generation (Fig. 3.5).

Nowadays, it is widely recognized that important properties like the stiffness and strength are governed by processes that occur at one to several scales below the level of macroscopic observation.

In particular, adaptive materials structure functional analysis done in (Pleskachevsky et al 1999), foresees macroscopic, mesoscopic and microscopic levels of its detailing (Fig. 3.6).

Mesoelements of natural and artificial composites have dimensions of nanometers, micrometers (ultrafine fillers, macromolecules of polymers, biological cells) and millimeters (fibers, pores and films). Discrete methods taking into account the real geometry and deformation characteristics of mesoparticles allow adequately approximate the stress-strain state of the

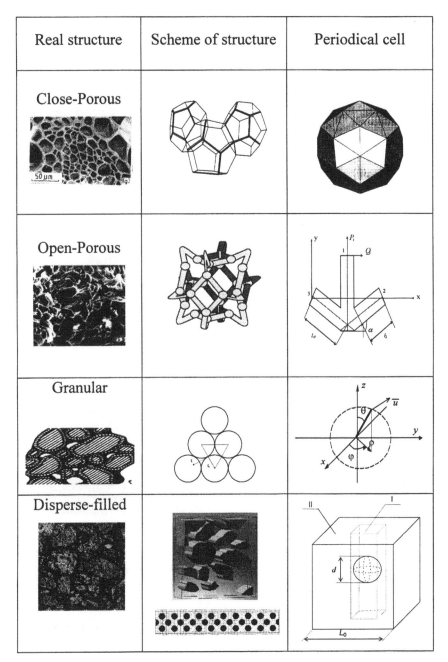

Fig. 3.3. Mesostructures and structural elements used in computer design of materials

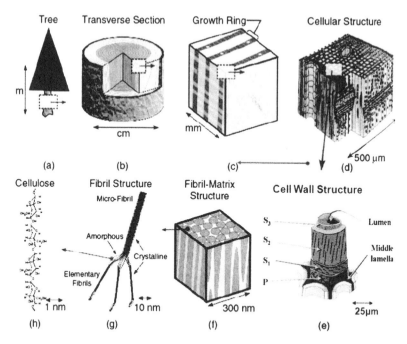

Fig. 3.4. Structure hierarchy of wood (Moon 2011)

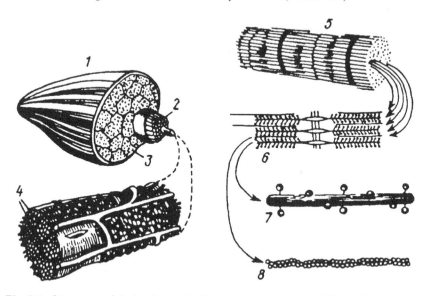

Fig. 3.5. Structure of skeletal muscle: *1* – muscle; *2* – group of fibers; *3* – fascia; *4* – fiber; *5* – monofiber; *6* – sarcomer; *7* – miozyne thread; *8* – actyne thread

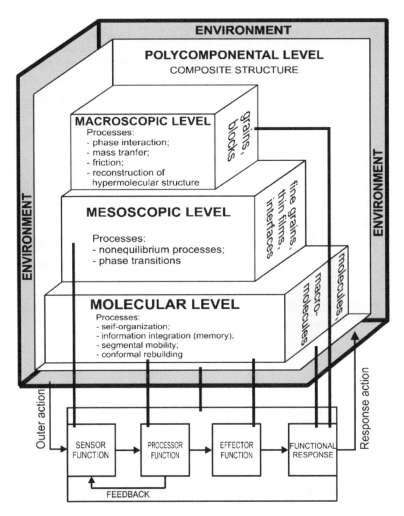

Fig. 3.6. Differentiation of structural levels of adaptive materials

composite, upon which perform the strength and deformability design of the smart materials. It is possible to formulate the statement and the solution of direct and inverse problems. The formulation of direct task (that is much more common) consists of the composite material characteristics calculation according to the available experimental data for components. More complex inverse problem of material implies reconstruction of "internal" parameters (Anokhina et al 2010) or design of composite structure, providing the implementation of the desired properties (Lyukshin et al 2011). Multi-level method proposed in (Shil'ko, et al 2013a, Shil'ko et al 2013b) illustrates the solution of this problem (Fig. 3.7).

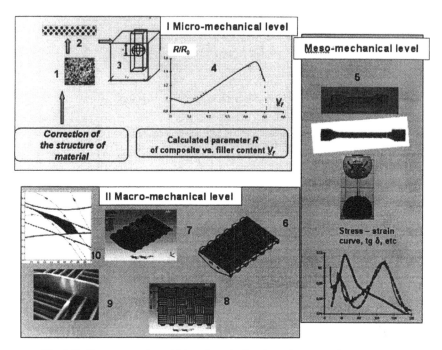

Fig. 3.7. Sequence of multi-level optimization of composite structure

3.2.4. SM analysis as mechanical modeling of composites with movable interface boundaries

A certain progress in the structural and functional analysis of composites has been achieved through the concepts of fractality (fractals – are sets with extremely irregularly branched or jagged structure) (Ivanova et al 1994). The fractality reflects not only the morphology, but also the evolution of the object's structure, with a fractional dimension greater than the topological one. A multifractal description provides an additional opportunity for the analysis of materials' mechanical properties that allows entering quantitative indicators of homogeneity and order for the different structures. Within the scope of the multifractal formalism, on the basis of the composite structural characteristics only, it is possible to form digital diagrams with the parameters that determine the properties of the composite material.

"Memory" of materials (for example, "shape memory") is a specific property of the SM to store the information about the history of previous technological and operational impacts. The FBS reactions, therefore, are committed considering the predecessor events. The "shape memory" effect (Wei et al 1998, Sun et al 2012), in this regard, serves not only as driving force of self-organization, but also – as a mean for recording the material loading history. The SM mechanical response for an external action contains a specific nonlinearity that is associated with the existence of the FBS channel by which a material self-organization takes place. Analytical methods for this nonlinearity description are seemed ineffective. It is desirable to use discrete models and algorithms of control theory, permitting to obtain the required parameters distribution in a finite number of iterations without restrictions to the interphase and external boundaries configuration.

This brief analysis shows that the mechanics of materials is sensitive to changes in the complexity level of the object study. During creation of traditional simple (mainly homogeneous) and composite materials, their desired properties are made real at the manufacturing stage. As a rule, such systems, within their functioning, deteriorate with some kind of speed at uncontrolled processes of creep, continuity damage and other processes leading to the deterioration of their service characteristics and availability loss.

Since the SM development presupposes creation of self-organization mechanism of a structure causing the total material energy decrease, the assessment of its thermodynamic state is very important. This estimation process includes analysis of the phenomena as chaos, stabilization, cooperative movement of the material particles, bifurcation points determination, cooperative effects, etc., and may be found in several studies. Such classic examples of structure self-organization as plastic deformation (Makushok 1988) and transferred films formation during friction (Gershman 2006, Bushe and Gershman 2006) are taken here with regard to the mechanics of materials.

The modeling of SM structure self-organization may be based on the cellular automata theory (Toffoli and Margolus 1987). The cellular automata or "computing space" with the parallel chains architecture simulates the parallelism of any material structure. In the context of discrete time, the reaction of such system corresponds to a typical set of rules by which a cell at each step computes its new state according to the states of the neighbors. The cellular automata allows (by force of numerical experiments) to establish the dependence of the cells' cooperative "behavior" on the regularities of a single cell (mesofragment of material) evolution. However, the hypothesis of the cells' aggregate non-deformability used in the classical method of cellular automata, imposes restrictions on the application of this method for kinetic phenomena modeling. In conclusion, it is necessary to note that the concept of "accommodation" that characterizes adaptive materials is not equivalent to the concept of "self-organization" inherent in the SM.

As for the modeling of structure self-organization, it is important to mention the role of the interphase boundaries which largely determine the composite material properties. Moreover, the ability to change the configuration of these interfaces creates the preconditions for the composite properties effective control.

The use of mesomechanical concepts to describe the SM with free (movable) interface boundaries has become a pressing problem. The simplest composite SM scheme is shown in Fig. 3.8, where the changes of the material's effective characteristics occur as a result of changes in the configuration of the components interface.

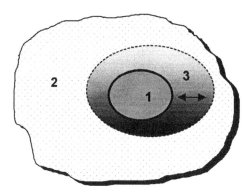

Fig. 3.8. Structural element's scheme of composite SM: filler particle (*1*), matrix (*2*), interface (*3*)

Variational methods of mechanics are used to describe the moving boundaries of SM structure adapting to the external action. The variational principle or an empirically established regularity (e.g. the Wolff's law of tissue strength balance) may be taken as a criterion for the formation of *a priori* unknown and evolving SM regions. Theory of variational inequalities (Duvaut and Lions 1972) was used for interpretation of self-locking effect at friction (Shil'ko 1995). The study of moving outside and inside contacting boundaries under indentation of the granular material fragment with adhesive and frictional bonds may be considered as an example of SM description (so called "model of wet sand") (Fig. 3.9).

In the process of a fragment structure (macroscopic level) loading, specific zones coexist at the adjacent granules contacts (mesolevel) with boundary conditions corresponding to the state of the contact – separation, adhesion and slippage. Acts of the adhesive bonds destruction occur when the normal p and tangential τ stresses reach critical values in separation and shear, respectively. Thus, the problem of the non-limiting equilibrium of a system with constraints in the form of inequalities

$$|\tau| \leq \tau_s, \ |\tau| \leq \tau_a, \ p \leq p_a, \tag{3.2}$$

Fig. 3.9. Stress state of granular material under contact loading

where τ_s and τ_a are the yield stress and the shear strength, p_a – the adhesion strength) is formulated for the mechanics. Compliant with the principle of the energy functional I minimum, state of the adjacent granules contact (at kinematically admissible displacements v) is described by the equation (Duvaut and Lions 1972)

$$I(u) = \min I(v), \tag{3.3}$$

$$I(v) = \frac{1}{2} a(v,\,v) - L(v) + \int_S \tau |v_T|\, ds, \tag{3.4}$$

where u is exact solution of equation $I(v)$, $L(v) = \int F v\, ds$ is linear part, $a(v,\,v) = \int a_{ijkl}\,\varepsilon_{kl}(v)\,\varepsilon_{ij}(v)\, ds$ is quadratic part of the functional, S is contact's surface area.

The assumption about the reversibility of the destruction process (adhesive bonds recombination) is accounted for in the modified variational inequality (Shil'ko and Pleskachevsky 2001)

$$a\left(u^{t+dt},\, w - u^{t+dt}\right) - L_*^{t+dt}\left(w - u^{t+dt}\right) + j^{t+dt}\left(w - u^{t+dt}\right) \geq 0, \tag{3.5}$$

where j^{t+dt} component describes the adhesive bonds destruction

$$j^{t+dt}\left(w - u^{t+dt}\right) = \int_S \left[\tau^t + \left(\tau_s - \tau_a\right)\theta_0\left(\tau_a - \tau_s\right)\right]\left(|w_T - u_T|\right) ds, \tag{3.6}$$

where θ_0 is the Heaviside function. The possibility of their recovery corresponds to the expression

$$\tau_a^{t+dt} = \tau_a^t - k_\tau(t)\,\tau_a^t\,\theta_0\left[\left(u_T^{t+dt} - u_T^t\right)/ dt\right]. \tag{3.7}$$

The stress field calculation stepping in time *t* was done to evaluate the microcontacts current state. Computer experiment showed that the initial state of the granular material has a stress concentration (see Fig. 3.9). The acts of separation and shear, in the presence of ruined relationships competing recombination process, provide material's continuousness saving without increasing of the stress concentration and without fracture during the load growth, as it is in the case of elastoplastic deformation of conventional materials. Therefore, the creation of "smart sand" without stress concentration zones adds up to optimal control of the moving interface boundaries.

The peculiarity of SM mechanics consists in consideration not of a single fixed material structure that cannot be optimal for a wide range of operational effects in principle, but flexible structure that in initial state is not only optimal, but a capable of reasonable modification. Thus, the SM development supposes the mobility of basic structure and formulation of material function criterion that regulates target oriented change of its properties and, consequently, the state equation if situation changes. For example, the materials show signs of artificial intelligence due to the non-destruction criterion if, adjusting themselves to mechanical loading, they change their structure (as in the case of the "smart sand") and implement the strength balance function programmed by a developer.

The composite's smart response to the external effect can be realized through physical, chemical and biological mechanisms, it was focused in Chapter 2. In principle, SM does not necessarily need a heterogeneous structure, since it is possible to provide an adaptive response, for example, by changing the shape of the article. However, the number of potential ways to implement artificial intelligence in the multiphase heterogeneous materials is immeasurably greater than in homogeneous ones. As a result, it can be insisted that composites are the SM immediate precursors.

Let us assume the classification of materials with account of interrelations found between structure and functions as well as analysis and modeling of intellectual systems. Some of the assumptions put forward by the authors are based on the theory of functional systems and synergism (Prigogine and Stengers 1984). Three generations of materials which can be discriminated in the proposed classification (Shil'ko and Pleskachevsky 2003), are given in Tables 3.3 and 3.4.

The first generation of traditional materials is formed including monofunctional ones whose properties are determined by the nature and initial quality of a single component. The next are traditional composites with a prominent structural hierarchy, also being monofunctional. They are characterized by stability of inner and external boundaries, i.e. fixed structure of components, intermediate layers and the composite as a whole.

Table 3.3 Evolution of structure and properties of materials

Criteria of material quality	Characteristics of different generations of materials		
	Simple material	Composite material	SM
Function and structure	Monofunctional single-component material	Polyfunctional polycomponental material with fixed boundaries between components	Polyfunctional polycomponental material with movable boundaries between components
Optimality degree	Initial property of monocomponent	Initial property of components and interphase layers	Efficiency of sensing extreme effects and elimination of refusals
Method of properties controlling	Physical and chemical modification of origin properties	Properties are efficiently regulated technologically based on principles of adaptation and synergism	Self-regulation of structure based on feedback system as a rational reaction on external conditions

Smart materials with coordinated functions, active and adaptive behavior belong to the third advanced generation of materials. They are more intelligent than ordinary materials because perceive outer effects at unchanged function owing to, presumably, structural self-organization. In this connection, the mobility of the component boundaries should be remembered as an indispensable property of smart materials, which is not present in traditional composites.

The qualitative transition of materials from the passive to an active functioning is shown in Table 3.4. Naturally, prerequisites of such a transition are formed at the levels of two preceding generations. Thus, transformation of one physical field into another (e.g., piezo- or photo effects) is probable at the stage of monofunctional material. The creation of qualitatively new (emerged) state, including forecast properties, is a logical continuation of the additive and synergetic principles of composite production. This precedes the development of adaptive composites, being a subclass of smart systems with the dominating accommodation strategy.

The suggested classification makes it possible to forecast other unknown materials of the intellectual type, for example, capable of self-destruction "kamikaze", those ensuring partial or full restoration "regenerators" and materials offering programmed control of the environment ("cyber") and implicit ("incognito") ones. These subclasses constitute a

Table 3.4. Systematization of materials by general criteria

Functional evolution	Degree of activity	Degree of intellect	Functioning quality	Mode of behavior
mono-functional	passive	'trivial'	material	'forecast'
	active	'wit' (functional)		
poly-functional	active	smart (adaptive)	material = part	'indefinite'
				'egoist'
			material = system	'time-server'
		'wise' (ecophilous)	material = medium	'kamikaze'
				'regenerate'
				'cyber'
				'incognito'

new type of "ecophilous" materials which behavior supports homeostasis of the environment.

Relative simplicity of **adaptive composites** is due to their orientation aimed to fulfill only the accommodation function of the part or a system in contrast to a higher status of the material-medium subclass (Table 3.4). However the adaptive composite is formed rather in time than by a mechanical mixing of structural components, and evolves as a specific unit by coordinating interrelated physical processes based on an imparted optimum criterion. In this case, the emergence of macrostructure is specified by origination of collective modes under the action of fluctuations, there competing and, finally, by selection of the most accommodated mode or their combination (Prigogine and Stengers, 1984). The structures themselves could be described in physical terms as types of adaptation to outer conditions.

The following scheme of smart reaction of material, based on the principle of interface boundaries mobility, is presented in the paper (Shil'ko and Pleskachevsky 2003):

- the interface boundaries' configuration, that defines physical and mechanical characteristics of the material, depends on the external action;
- the material self-organization at the micro level occurs by a spontaneous alteration of movable interfaces and by establishing optimal

values of the material's physical and mechanical characteristics in bifurcation points;

- specified by developer's material smart response to the external action by use of the FBS which regulates the bifurcation points parameters.

3.2.5 Self-organization of material structure

Reaction of a material due to mutual coordination of structural and functional parameters of microsystems characterizes it as an open self-regulating system. Selection of the mode of behavior in response to outer effect does not arise from the principle of the least action, neither from the principle of compulsion (Gauss principle) nor from that of the utmost probability. Active response systems eliminate (or subordinate) contingency. This makes grounds to speak about a programmed behavior of the system, i.e. the decision is made according to the inner criteria determined by the structure itself and system parameters, which substantiates the necessity of direct and reverse connection channels.

It follows from the above that to form a more complex processor function of SM it is possible to use the universe phenomenon of self-organization, which is not limited to only systems of higher organization and functional complexity and isn't a monopoly of bio- or social systems. A self-organizing system is understood as a system capable of stabilizing parameters under varying outer conditions through directed ordering of its structural and functional relations aimed at withstanding entropic factors of the environment, which helps to preserve its characteristics as an integral formation.

The material formed by combining its components acquires the characteristics of a composite structure, which is a notion nonequivalent to the structure of its constituents. This fact raises composite materials to a higher structural level and admits the probability of per layer differentiation of the functions in order to reach the integral control system (Prigogine and Stengers 1984). In our view, to realize adaptation mechanism to outer conditions in composite materials, it's worthwhile considering the combination of different scale physical processes, where we single out at least 4 structural levels: molecular, mesoscopic, macroscopic and polycomponental (Fig. 3.6).

The molecular level is the basic one at programming material behavior. This is because its scale in polymer composites corresponds to cooperative effects of segmental mobility and conformal rebuilding that provides conditions for self-organization in high-molecular bodies (Mashkov et al 2013). Just here the processor function is realized as a capacity for estimating variations due to outer effects and as a tool formulating the character and force of response based on stationary characteristics of the microsystem. Also, the effector function is fulfilled here for exciting reverse reactions by varying characteristics of the microsystem on a self-organization base.

The mesoscopic level performs the sensor function as an ability to perceive outer effects. Non-equilibrium processes are initiated at this level changing molecular structure and supporting the interaction of direct and reverse channels between the levels.

The macroscopic level makes provision for the mobile function as a reorganization of the initial subsystems (components) aimed at preserving the behavior model.

The mobile function is also realized at the polycomponental level, though intention in this case to provide the system (material = article) functioning as a whole.

To organize control, the processes relating to the mention levels should be coordinated using functional links between them.

It is to be remembered that polymer composites are potential carriers of intellectual properties. Namely, they are sensitive to physical fields, i.e. show a sensor function; make it possible to carry out the actuator function (shape memory of thermosetting resins, etc) and, finally, among all other artificial material media they most closely approach the living nature (biotissues are usually built of high-molecular compounds).

The study of synergetic phenomena in nonliving nature as a linking element between analogous processes in original objects will, in our opinion, provide a possibility to find structural-and-functional bioprototypes of artificial SM.

3.3 The materials with 'negative' characteristics as source of smart effects in structures: Auxetics

3.3.1 What are auxetics?

The effort towards improving the performance of novel devices based upon realization of non-linear and non-trivial (smart) deformation properties of materials is the aim of many current investigations. First, we shall consider the materials with a negative Poisson's ratio, v, termed 'auxetics' (Evans 1991).

Data structuring, examination of the mechanisms of generating the negative Poisson's ratio and analysis of likely applications for auxetics have been discussed in (Wojciechowski et al 2007) and reviewed particularly in (Koniok et al 2004, Yanping and Hong 2010). Poisson's ratio affects a very important mechanical property, i.e. compressibility and shear stiffness of a material. Under a uniaxial stress, auxetics expand/contract at the direction perpendicular to the tension/compression direction, respectively as shown in Fig. 3.10.

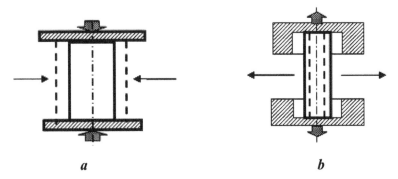

a	*b*

Fig. 3.10. The deformation mode of an auxetic material under uniaxial stress: *a* – compression, *b* – tension. Initial configuration before loading has been shown by dashed lines

This property should influence stiffness and slip under contact loading, and in this way allow control over deformability and friction characteristics of composites and tribojoints based on auxetics. As will be shown, the contact characteristics vary dramatically with variation of the sign of Poisson's ratio. In the classical elasticity theory for isotropic bodies (Love 1944, Landau and Lifshits 1986) Poisson's ratio $\nu = (3K - 2\mu)/(6K + 2\mu)$, where μ, and K are the shear and volume moduli respectively, the Poisson's ratio of isotropic bodies can vary in the limits $-1 \leq \nu \leq 0.5$. The upper limit corresponds to incompressible materials, e.g. rubber, whose volume remains constant at significant shape variations; the lower one belongs to the materials preserving their geometrical form with changing volume.

Several natural and artificial auxetic materials have been described to date, but experimental and theoretical studies of the smart frictional and mechanical properties of these materials are still not well developed. For example, the possibility for realization of self-locking effect in contact joints containing auxetic components will be shown in p. 3.4.3.

The approaches available for creation of composites with $\nu < 0$ assume either the use of individual auxetic components or formation of an auxetic composite – a combination of special structural units of mesoscopic level (pores, granules, crystals, permolecular formations of polymers, etc).

3.3.2. Porous and granulous auxetics

Porous (cellular) materials like 'solid – gas' inhomogeneous systems are efficient structures in respect to optimizing strength and stiffness for a given weight. So, according to estimation, different porous biological tissues as, for example, human's cortical bones may show the auxetic properties and it was important to verify such behavior experimentally (Overaker et al 1999).

These materials are useful for cushioning, insulating, damping, absorbing the kinetic energy from impact, packing, etc. Stiff and strong ones are preferable in load-bearing structures such as a lightweight core in sandwich panels. The term 'cellular' is appropriate when the material contains polyhedral closed cells, as if it had resulted from solidification of liquid foam. Auxetic porous materials, including auxetic porous nanomaterials, having very high mechanical properties, are suitable for creating adaptive contact joints and for replacing natural materials such as damaged bone and tooth biotissues. A successful combination of porous material customer values' (low density, workability, insulating) and highly developed theory of the cellular structures (Hilyard and Cunningham 1994, Gibson and Ashby 1997) gives rise to the increased interest in this specified type of auxetics mentioned. The cellular plastic structure is replaced by a rod-shaped design with the cells in the form of regular polyhedra, calculating the material effective characteristics. The open-cell two-dimensional honeycomb structure, as it is proposed in (Kolpakov 1985), is used to analyze anisotropic auxetic materials of low density with open porous and separation of a periodicity cell consisting of three rods connected in a rigid assembly.

The all-round cellular plastics (open-cell foam plastic, porous foam) were exposed to a compression and heating to a softening temperature with a subsequent cooling to the ambient temperature is the main methods for the producing of the auxetic thermoplastic porous materials (Lakes 1987). The polyester, polyurethane, polyethylene and copper sponge based foams were subjected to such heat treatment.

Transformation of a conventional cellular plastic in auxetics such way is possible with the degree V_{in}/V_{tr} of volume compression equals to $1.4 \div 4.8$, where V_{in} and V_{tr} are properly the initial and transformed material volumes. The analysis of the cellular plastics microstructure shows that the ribs of pours bend inwards (Fig. 3.11), so under the stretching / compression the sample cross section is increased / decreased.

The ν values decrease up to -0.7 with an increase in volume compression (Lakes 1987). The cells collapsing begins at large compression deformations ($V_{in} / V_{tr} > 3.5$), and transformation is impossible when $V_{in} / V_{tr} > 5$.

A method (process) for the production of auxetic materials during foaming stage without use of an additional heating and of providing triaxial compression devices was proposed later (Konyok et al 2004).

The auxetic porous materials based on the polytetrafluoroethylene (PTFE) with a pore size of less than 150 μm have very low values of $\nu \approx$

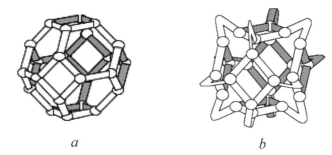

a   *b*

Fig. 3.11. The structure of the initial (*a*) and auxetic (*b*) porous materials

–12 at 15 % relative deformations (Evans and Caddok 1989). The samples after special treatment were subjected to a high-speed orientation drawing (elongation) during heating and drying.

The microporous auxetics (Alderson and Evans 1992) are obtained by compressing heated to 160°C ultra-high-molecular polyethylene (UHMPE) powder and its twin extruding through the holes of 5 and 15 mm diameter. The extrudate's Poisson's ratio was defined at the compression deformation range of the 2÷3 % and it reaches the value –6 for the low modulus and –1.5 for high modulus (0.2 GPa) samples. The microstructure's analysis shows that the material consists of fibrils connected quasi-spherical particles with diameters of 20÷30 μm. The structure's deformation is accompanied by a competing process of the fibrils transverse displacement and nodes rotation (Evans and Caddok 1989).

One can consider the granulous materials as a kind of porous structure. Firstly, Poisson himself came out with a suggestion that there is a relationship between the v coefficient and microstructure of granulous materials. He took it for granted that all the materials consisted of spherical smooth particles randomly packed inside and interacting with one another along the directions connecting their centers. He set up such that systems had $v = 1/4$.

Poisson's ratio depends on the ratio of normal and tangential stiffness of the particles contact $\lambda = k_t/k_n$: in the two-dimensional working example $v = (1 - \lambda)/(3 + \lambda)$ and in three dimensional case $v = (1-\lambda)/(4+\lambda)$ (Rothenburg et al 1991). The parameter $\lambda = 0$ for the smooth spheres having no shear resistance and $v = 1/4$ in this pattern, this model coincides with the results obtained by Poisson but, when the tangential stiffness exceeds the normal stiffness, values of v are negative if $\lambda > 1$.

3.3.3 Crystal auxetics

Monocrystals are axially auxetic if negative values of v become obvious alongside crystallographic directions <100>. If the negative v values are observed along the crystallographic directions that do not coincide with the <100> directions the monocrystals are non-axially auxetic. As a rule, the face-centered cubic phases (fcc) and body-centered cubic (bcc) phases of elementary metals show signs of non-axially auxetic properties. The auxetic crystals criterion: $S_{11}+S_{33}+2S_{13}-S_{44}>0$, where S_{ij} are the components of the crystal elastic compliance matrix, was proposed with the intention to simplify the search of phases having negative ν values for at least one of the crystallographic directions (Svetlov et al 1988).

A model was composed to show that the extreme ν values for the bcc and fcc crystal lattices are equal to 1 and 2 and 0 and + 1/2 in that order (Baughman et al 1998). Thus, in bcc crystals there is a plane parallel to the axis of elongation and reducing its area under tension (Fig. 3.12).

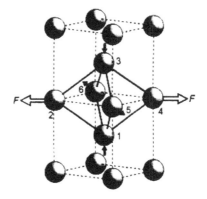

Fig. 3.12. Structural origin of Poisson's ratio anomalous values for the bcc crystal with hard spheres (Svetlov et al 1988). White arrows indicate the forces directions, the black ones – shifts of atoms 1÷6 under the force F applied along the <110> direction

The auxeticity was initially discovered among the allomorphic SiO_2 tetragonal monocrystals of α-cristobalite. A model, where the structure of the crystal is represented as a number of interlinked SiO_4 tetrahedra simultaneously rotating and changing in their size during a uniaxial deformation, was proposed to explain the mechanism of this effect (Alderson and Evans. 2001). Moreover, the rhombohedral crystals of α-quartz exhibit auxeticity if their uniaxial deformations is more than 5 %.

The results of the calculations carried out for the fcc and hexagonal close-packed cells, have shown that the limit ν value for these crystals is 0.33 (Theocaris 1994). The same calculations done for the quasi-isotropic polycrystalline α-cristobalite aggregates while using the averages of Voigt and Reus nevertheless, give values of ν equal to –0.13 and –0.19 accordingly.

A number of auxetic single crystals (α-cristobalite, pyrite) are natural minerals (Alderson and Evans 2001). There are reports that porous sandstones have $\nu < 0$ (Grima et al. 2000). The compressibility of natural zeolites has been studied because many of zeolites have a structure similar to α-cristobalite crystals.

The model calculations of Poisson's ratio for some aluminosilicates and aluminophosphates has been done with the molecular modeling method and it allows to predict the value $\nu < 0$ for single crystals of $Al_{20}Si_{20}O_{80}$ and $Al_{16}Si_{24}O_{80}$ along (001) plane, and for single crystals of $Al_{36}P_{36}O_{144}$ along (100) plane. Negative ν values are possible for the polycrystalline aggregates thereof.

The viruses' agglomerates classified as colloidal crystals class with body-centered cubic lattice are exotic example of auxetics (Schmidt et al 1993).

3.3.4. Composite auxetic materials

The auxeticity phenomena are noted in the natural composites. Poisson's ratios calculation based on the experimentally determined elasticity moduli of the strongly anisotropic birch wood and of the coniferous species wood showed that ν assumes negative values while the direction of stretching was 45° to the line laying fibers (Ashkenazi and Ganov 1980). The erythrocytes (red blood cells) deformation is determined mainly by conformation of their external protein skeleton, thus they may be placed among auxetics (Schmidt et al. 1993).

Theoretical and experimental approach to the creation of artificial composites with $\nu < 0$ implies the joining and combination of the auxetic phases or obtaining of the auxetic materials in the form of mesostructure consisting of units joint in a certain manner and equivalent to the continuous medium volume.

The development of quasi-isotropic composites filled with the *auxetics particles* draw author's attention as it provides obtaining of technological and low-cost materials with $\nu < 0$. Calculations of the composite with spherical and ellipsoidal inclusions effective characteristics for the two-dimensional and three-dimensional matrices are done on the assumption of random location the filler material particles in these matrices (Wei and Edvards 1999). The simulation results with different ratios δ of hardness for the filler and the matrix in view of that inclusion fractions volume is 45% are shown in Table 3.4.

The filler concentrations where above the composite becomes auxetics are found.

The $\nu_c(\delta)$ dependence analysis shows that for quantities of δ, where $0 < \delta < 1$, parameter ν_c decreases if δ increases, reaching its minimum and

Table 3.5. Effective Poisson's ratio v_c of composite at $v = -0.9$.

Form of inclusion	v_c values for various quantities of δ		
	0,1	1,0	10
Disc (2D)	−0.3020	−0.2856	0.1216
Disc (3D)	−0.0385	−0.3575	−0.7387
Sphere	−0.0624	−0.2081	0.0650
Wedge (2D)	−0.2679	−0.2266	−0.0508
Needle (3D)	−0.0555	−0.1714	−0.0562

followed with a monotonic increase and access to a horizontal plateau for the spherical, needle-shaped and disc inclusions.

The analysis of *auxetic composites with non-auxetic inclusions* will be carried out at three levels:

- macroscopic level, where the occurrence of a negative Poisson's ratio can be explained by the theory of anisotropic bodies' elasticity in the representation of the composite as a continuous medium;
- mesoscopic level, when auxetic properties is determined by way of material structural units in the form of cells, pours, and other inclusions (Milton 1992);
- microscopic level, calls forth materials' elasticity with in the intra- and intermolecular interactions.

For the composite formed by the bevelled fiber laying in an elastic incompressible matrix (Fig 3.13, *a*), the following values of Poisson's ratio have received in terms of **macroscopic** theory (Clark 1963, Akasaka 1989)

$$v_{xz} = 1 - \left(E_1 / 2 + E_2 \sin^2 \theta \cos^2 \theta\right) / \left(E_1 + E_2 \sin^4 \theta\right). \qquad (3.8)$$

When $E_2 \ll E_1$ E_1 we have $v_{xz} \approx 1 - ctg^2\theta$, where E_1 and E_2 are Young's moduli of the matrix and of the fiber.

Poisson's ratio v_{xz} takes negative values for the small angles θ of fiber's laying. The model's *a* deformation conduces to the changes in the fibers orientation. These fibers slightly elongates if to compare with the low-modulus matrix and promote its pressing-out perpendicular to the plane of the reinforcement. Under the small angles of fiber laying (10° – 40°) in the laminates formed on the carbon fibers with an epoxy binder prepregs basis (Peel 2007) v_{xz} is extremely low, less than −(50–60)!

G. Milton has proposed another way to 'design' a composite material where $v \sim -1$ basing on the mesomechanical approach (Milton 1992). The Fig. 3.13, *b* shows an idea of creating a two-phase isotropic composite with

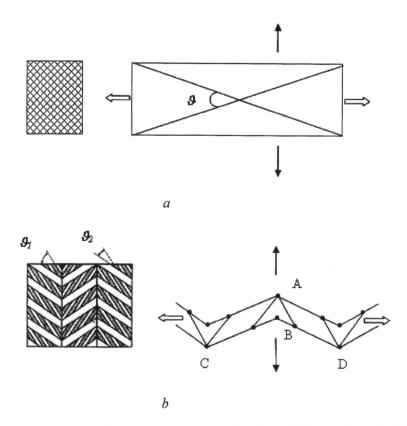

Fig. 3.13. Structure and models of anisotropic (*a*) (Clark 1963) and isotropic (*b*) (Milton 1992) auxetic composites; white arrows indicate the direction of forces, black – displacements.

$\nu \sim -1$ on the basis of hard fibers and compliant matrix, where the shaded areas correspond to the compliant phase and are modeled with extensible but unbending rods. The AB size increasing is directly proportional to an increase in the CD length under an infinitesimal deformation of the rod model itself. If you combine a few of these structures, the resulting 'stack' will simulate the auxetic laminate.

Calculation of the auxetic composite with a star shape of its inclusions effective elasticity (Theocaris et al 1997) by the numerical homogenization method demonstrated that negative ν values occur under the cell concavity angles of 36° – 89°. Most of all, the effective value of Poisson's ratio in this family of composites is determined with the geometry of inclusions and less with the ratio of inclusion and matrix stiffness.

The possibility of creating auxetics based on the filled polymers is analyzed on microscopic level (Kolupaev et al 1996). The composites based on thermoplastic polyurethane filled with ultrafine (0.3 ÷ 1.0 μm) particles of tungsten, iron or molybdenum, where $v \approx -(0.2 \div 0.4)$ were obtained. It was established that in the stress range 0.97 MPa $< \sigma <$ 7.11 MPa created by the inclusions in a matrix, the composite under consideration exhibits its auxeticity. It seems that it is possible to create SM on these principles.

3.3.5. Molecular and nanosize auxetic structures.

The Poisson's ratio negative effect implementation on a microscopic (molecular) level can radically improve the homogeneity and reduce deficiency of materials.

It is possible to transfer the principle of rigid and flexible elements combination implemented at the macro- and mesolevels in composites and poromaterials to the molecular level. It is shown with the help of the molecular mechanics methods (Evans et al. 1991, Grima and Evans 2000) that $v < 0$ observed in the polymer having the fragment of macromolecule corresponding to the cell exposed in Fig. 3.14.

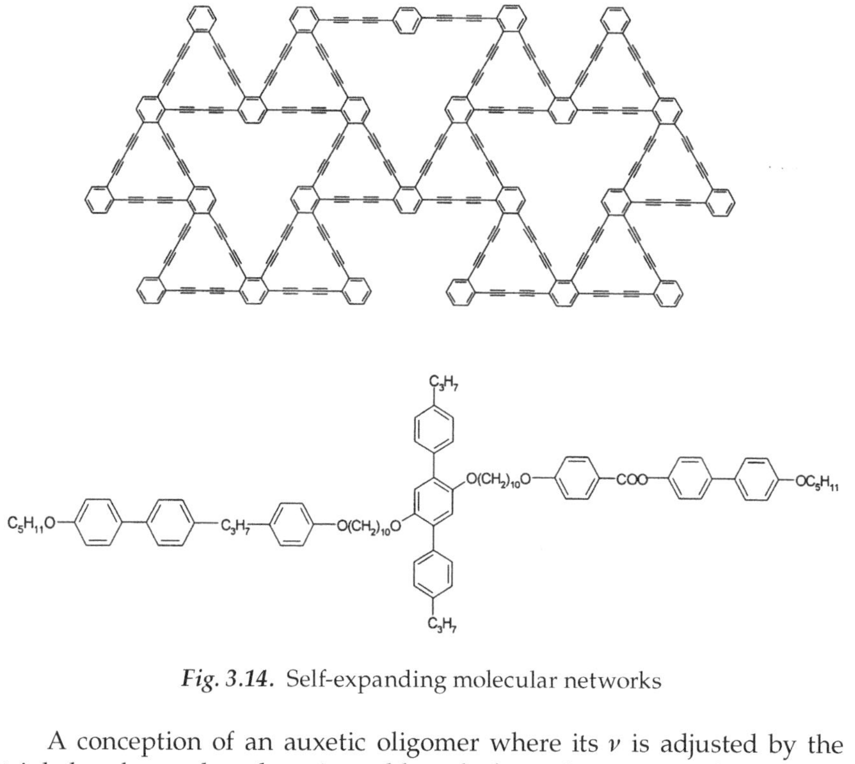

Fig. 3.14. Self-expanding molecular networks

A conception of an auxetic oligomer where its v is adjusted by the triple bonds number changing, although the real structure of molecular will be distorted in factual synthesis due to oxidation and its elastic properties will be affected the way of molecules "stacking", is envisaged. The calculations presented reveal feasibility of the negative ν values realization for the carbon type crystal structures of the cis- and trans-polydiacetylene with the main chain structure $(-C=C-C=C-)_n$ (Baughman and Galvao 1993).

This effect can be manifested in nematic liquid crystals where the axes of the molecules are oriented parallel to each other. If the general structure of the polymer's nematic phase is presented in the form of rigid rods connected by elastic threads, this polymer must have a certain free volume to obtain negative values of Poisson's ratio ν in the stretching. The polymers having strong intermolecular interaction must have smaller values of ν *ceteris paribus* in view of the fact that the value of ν depends on the lattice parameters and the free fluctuation volume.

The design of the three-dimensional auxetic molecules is a rather complex task. In the three-dimensional case there is no symmetry of the lattice providing elastic isotropy as in the two dimensional case, where for the formation of an isotropic periodic lattice the symmetry axis of the third order is quite enough. It is possible to obtain particles of SiO_2, for example, having stable structures as for α-cristobalite with an average negative ν value. The molecules capable to form auxetic elastically isotropic phases in the three dimensions have not been described in the literature yet. Nevertheless, chemistry demonstrates its rising opportunities and promises to discover similar structures in the future.

The auxetics were offered in the form of sheets consisting of single-wall and multiwall nanotubes. The properties of such sheets are strongly dependent on the content of nanotubes of both types. Sheets consisting mainly of single-walled nanotubes are deformed by stretching in a standard manner. The increase of multiwall nanotubes in number leads under certain conditions to the abrupt transformation of material into an auxetic.

3.4   *The statements and solutions of some SM mechanics problems*

Let us consider the examples of analysis and prediction of the composites smart reactions which are particularly add up to optimal localization of the movable interfaces. A number of SM mechanics problems are already formulated, namely those dealing with cracks self-healing, self-reinforcing, compensation of thermal expansion and self-assembling. These models are given below.

3.4.1 Self-healing of cracks

Destruction of materials as a result of mechanical loading or corrosion is accompanied with the advent and development of linear and volumetric microdefects – cracks, twins, crazes, etc. Till recently, the research in the field of the fracture mechanics and the physics of strength were aimed at predicting of the irreversible damage development outcome such as main crack (Bolotin 1984, Borst et al 1994, Morozov and Zernin 1999) that is a material bearing capacity loss of a forerunner.

The vital tissues demonstrate their ability to 'repair' damaged structures by self-healing at the initial microscopic level of material defects. The soft biological tissues (skin, cartilage, endothelium, etc.) as well as the hard ones (bone and dental) are the SM self-healing prototypes. Their lesions are 'repaired' as a result of the hemocoagulation (blood clotting) and the osteogenesis (bone cells expedient growth). Reestablishment of the technical materials continuity can be implemented on the basis of special physical and chemical processes proceeding (Ghosh 2009). Assessment of the self-healing materials working capacity is not only and not so much definition of their service life, but it is also the study of their adaptive possibilities, i.e. in comparison of the defect intensities' formation and the restorative reaction of opposite direction (competing reaction).

In SM, when linear crack achieves critical length, it 'triggers' the mechanism of diffusion growth inhibition – the reverse process of destruction. This 'smart' reaction is of a clearly defined threshold character, and in the process of material destruction the location and shape of the defect areas and main material itself are not known as well as their time-dependent behavior is unknown. Thus, materials self-healing modeling presupposes solving of the moving boundary mechanics problem.

But analytical and numerical modeling of self-healing materials is only in the initiation stage. For example, corresponding mechanical and mathematical description has been developed in terms of multiscale modeling (White et al 2006) as well as by means of continuum effective field modeling and lattice numerical simulations (Dementsov and Privman 2007).

Phenomenological statistical description of self-healing of cracks formed under the SM surface contact indentation is proposed in (Sergievich et al 2000). The expression for the probability density of cracks self-healing is obtained on the assumption that the statistics of linear defects (cracks) occurrence corresponds to the two-dimensional normal distribution, and their retardation – to Poisson distribution. Provided below an expression for the theoretical probability density for the defects mentioned occurrence in the point x takes into account the deformation history development and the crack length increment in the presence of the diffusion 'stoppers' in the material (Bashmakov et al. 1983):

$$p_k(x) = \frac{e^{-\lambda_k}\sqrt{|B^{-1}|}}{2\pi} \sum_{m=0}^{\infty} \frac{\lambda_k^m}{m!} \exp\left\{-\frac{1}{2}(x-(l_0-\lambda_k x^{(k)}) - mx^{(k)},\right.$$

$$\left. B^{-1}\left(x-(l_0-\lambda_k^{(k)}) - mx^{(k)}\right)\right\};$$

$$(3.9)$$

where $l_0 = \left(l_0^{(1)}; l_0^{(2)}\right) = \lim_{n\to\infty} \sum_s M\eta_{ns}$ is initial crack length; M is mathematical expectation of the crack length as of a random value; η_{ns} is crack length increment; n is quantity of the crack length increments; k is crack number; m is number of possible realizations of a random event; s is batch number of crack length increments; λ_k and B^{-1} are dispersion parameters of the normal and of Poisson distributions, respectively; $x(k)$ are coordinates of the k-th crack on the plane.

This expression is a composition of two distributions: two-dimension normal distribution with correlation matrix B and the Poisson distribution along the semidirect $x = ax(k)$, $0 \leq a < \infty$ with dispersion parameter λ_k.

The probability density for the collection of the star-positioned ray cracks (this type of breakage is observed at brittle materials indentation) is compiled as the sum of the densities in different directions $x = ax^{(k)}$, $k = \overline{1, v}$, multiplied by probabilistic weights q_k

$$p(x) = \sum_{k=1}^{v} q_k p_k(x), \qquad (3.10)$$

where v is the number of cracks; $\sum_{k=1}^{v} q_k = 1$, $q_k \geq 0$.

As an example, results of the material surface damage simulation under the action of a conical indenter point load are shown in Fig. 3.15. In the course of time, dispersion parameters of the probability densities where the expeditious interconnection of diffusion and destruction achieved are the SM decreasing quantity. The presence of the diffusion 'stoppers' in the SM as opposed to the usual one stipulates self-trapping of destruction initiated by a defect itself. Statistically, it is expressed as a negligible probability of a dangerous size crack emergence.

3.4.2 Self-reinforcing of multimodular materials

The theoretical foundation for the theory of adaptability for statistically indeterminate systems made of ideal elastoplastic materials were laid and in generalized view revealed in some publications in the middle of the last century. The analysis of the critical structures elastoplastic adaptability is

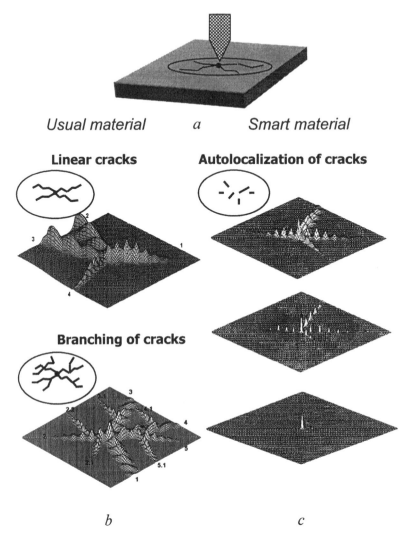

Fig. 3.15. Scheme of material sample point loading (*a*); schemes and computer images of distribution of linear and branching cracks probability density in the usual material (*b*) and in SM (*c*). Numbers on the images (*b*) correspond to the main cracks and their branching

accompanied with their strength calculation under the limit of load single unit action.

The idea of the elastic adaptability, or in other words – self-reinforcing of the multimodular strain sensing material, – was apparently voiced by for the first time in (Pleskachevsky et al 1999), and based at using the phenomena of multimodularity experimentally detected (Bell 1973), as well as very promising deformation effects named in (Hirotsu 1991, Wang et al 2007, Shang and Lake 2007, Lakes and Wojciechowski 2008) as 'negative bulk modulus', 'negative compressibility' and 'negative stiffness'. As a result of the elastic modulus automatic generation of piecewise inhomogeneous distribution, the comprehensive reduction of the stress concentrations may occur.

Assume the material possesses multimodularity and quantized transitions and value of Young's modulus are described by the operator E (Pleskachevsky et al 1999)

$$E = \begin{cases} E_1, & if \quad \sigma^0 \leq \sigma \leq \sigma^1; \\ \text{-- -- -- -- -- -- --} \\ E_l, & if \quad \sigma^{l-1} \leq \sigma \leq \sigma^l;, \\ \text{-- -- -- -- -- -- --} \\ E_n, & if \quad \sigma^{n-1} \leq \sigma \leq \sigma^n, \end{cases} \tag{3.11}$$

where σ is control stress; E_l and σ^l are Young's modulus value and stress threshold corresponding to the l-transition correspondingly; n is total number of transitions.

Reaching of the internal stress thresholds leads to the formation of a coherent interphase boundary (i.e. ideal contact boundary of the material sub-areas with different values of Young's modulus (Fig. 3.16).

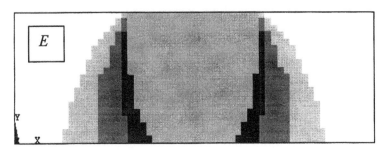

Fig. 3.16. Distribution of elastic regions in multimodular materials having 4 values of Young modulus E

The deformation of multimodularity material exhibiting two phase transitions, thus providing a variety of self-reinforcement alternatives, was examined. The material deformation diagrams described by a 4-parametric model are shown in Fig. 3.17. By setting the deformations and threshold (control) stress σ_y it is possible to determine these transitions, i.e. value of Young's modulus as a function of a load. Equivalent stress according to von Mises criterion is used as a control stress, and calculation of the stress-strain state components of the multimodular material is performed using finite elements method in the ANSYS software product.

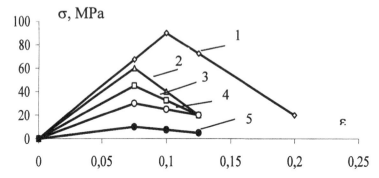

Fig. 3.17. The strain-stress curves for some multimodular materials: $1 - \sigma_1 = 90$ MPa, $\varepsilon_1 = 0,1$; $\sigma_2 = 20$ MPa, $\varepsilon_2 = 0,2$; $2 - \sigma_1 = 60$ MPa, $\varepsilon_1 = 0,075$; $\sigma_2 = 20$ MPa, $\varepsilon_2 = 0,125$; $3 - \sigma_1 = 45$ MPa, $\varepsilon_1 = 0,075$; $\sigma_2 = 20$ MPa, $\varepsilon_2 = 0,125$; $4 - \sigma_1 = 30$ MPa, $\varepsilon_1 = 0,075$; $\sigma_2 = 20$ MPa, $\varepsilon_2 = 0,125$; $5 - \sigma_1 = 10$ MPa, $\varepsilon_1 = 0,075$; $\sigma_2 = 5$ MPa, $\varepsilon_2 = 0,125$

As a preliminary, the task of assessing the equivalent stress within the material fragment given as thin plate having a square shape measuring 10x10 mm at an elongation value $u_y = -100$ µm (that is at 1% deformation, when the relations of Cauchy are true) was settled. A rigid fixation of the plate's lower edge provided the concentration stress areas development at the two corner points of the plate.

The degree of stress reduction concentration depends on the multimodularity parameters. Thus, variant 2 of multimodular material with parameters $\sigma_1 = 60$ MPa, $\varepsilon_1 = 0,075$; $\sigma_2 = 20$ MPa, $\varepsilon_2 = 0,125$ does not make noticeable reduction of the equivalent stresses near the fragment's fixed edge due to high values of the thresholds controlling stresses and deformations. The models 3-5 are more effective in the stress concentration reducing in this case.

After that, much more complicated problem about the stress conditions of the multimodular material in situation of contact deformation with friction was solved. The modeling of equivalent stress distribution in (Shil'ko 2011) demonstrate more expressed effect of this smart material adaptation to external load as shown on Fig. 3.18.

A rigid punch of square shape was indented into the multimodular material with Poisson's ratio $v = 0.4$.The punch dimensions were 5x5 mm, the coefficient of friction $f = 0.3$, Young modulus and Poisson's ratio of the punch material (steel) are equal to 210 GPa and 0.3 respectively. Normal punch displacement value was $u_y = -100$ µm, that was 1% of deformation

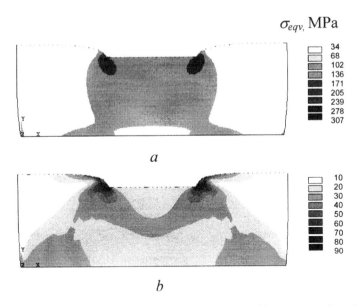

Fig. 3.18. Equivalent stress distribution in plate made of linear (*a*) and multimodular (*b*) materials under contact loading

ratio and, as in previously considered problem about tensile of the plate, corresponded to the equations of Cauchy.

The simulation results listed in Table 3.6 illustrate a reduction of contact and equivalent stresses and strain intensity and also with the proper selection of the parameters σ and ε the deformations according to von Mises.

Table 3.6 Comparison of stress-strain state of linear-elastic and multimodular materials at contact deformation

Material / Parameter	Linear-elastic	Multimodular (variants)				
		1	2	3	4	5
Contact stress, MPa	60.3	57.4	51.7	41.0	28.3	7.84
Tangential stress, MPa	17.2	12.8	11.6	9.01	6.35	1.67
Slippage, µm	1.90	2.60	2.57	2.48	2.74	2.89
Equivalent stress, MPa	20.3	17.1	22.8	16.8	11.3	3.62
Mises strain, %	3.20	2,67	2.39	1.76	1.18	0.38
Strain intensity, %	5.00	4.11	3.68	2.71	1.82	0.59

It is seen that multimodular material permits significantly to reduce a stress concentration due to the formation of sub-areas with two values Young's modulus. The most effective, among considered, is the model 5 providing an effective reduction of the maximal contact pressure (in 7.7 times), the tangential contact stress (by an order), the equivalent stress (in 5.6 times) and deformations (more than 8 times). Slippage in the contact does not change significantly (it is less than 40%).

When selecting multimodularity models for the specific conditions of the design loading, it is advisable to introduce additional criteria (strength hypotheses, stress components, elastic constants). The solution of similar problems for different stress concentrators (e.g. punches with a curvilinear boundary) and kinematic conditions causing constrained deformation is of particular interest. Analogy is observed here with the problems of geomechanics, where the phenomena of filtration, phase transitions during freezing, physical nonlinearity of soil, etc. are analyzed.

The idea caused to be attractive of SM obtaining out from the components with atypical (abnormal) intervals of changes in physical and mechanical properties caused by their unusual structure. The so-called inverted materials are among them, where physical constants as compared to conventional materials are reversed in the sign, for example materials with a negative coefficient of thermal expansion. Since the overwhelming majority of constructions operate under reversible elastic deformation conditions, the special elastic properties – 'shape memory', ultra-high compressibility, zero and negative thermal expansion, negative bulk modulus, etc. – are of considerable interest. Abnormally elastic materials constructed with the computer design methods (Askadskii and Kondrashchenko 1999, Ponte Castaneda et al 2004), are potential components of smart materials similar to their biological prototypes (Shil'ko 2011).

3.4.3. Self-reinforcing and damping of auxetic materials

The possibility of implementation the improved strength and deformation properties of composites and goods intensifies the materials with $v < 0$ research significance.

It is known that auxetic materials exhibit increased hardness (resistance to contact indentation) (Alderson et al 1994, Lakes and Elms 1993). Indentation of auxetic foams based on polyurethane and UHMPE as well as spongy copper showed dramatic growth yield compared to non-auxetic ones with the same porosity and apparent density. For UHMPE with $v = -0,8$ with the indenter load of 100 N, the indentation resistance increases to 8 times compared with the conventional porous polyethylene with $v \approx 0$.

Therefore, auxetics are promising as an intermediate layer of sandwich panels working under static and dynamic loads, as well as a means of increasing the contact stiffness for friction units.

Auxetic dunnage materials contribute to a more comfortable state of a man in the supine and sitting positions thus being a favorable body's weight distribution (Lakes and Lowe 2000); it is advantageous to use auxetic porous materials in artificial intervertebral disc (Martz et al 2005) as well as in joints' endoprosthesis design when it is necessary to provide a high anti-friction bearing capacity of the porous layer – a cartilage simulator (Shil'ko 2011).

Under a constrained deformation conditions with friction, auxetics exhibit the self-locking effect in friction joints (Shil'ko 1995, Shil'ko et al 2008). It is preferable to use an auxetic deformable element as a stopper of slippage in frictional joint (Fig. 3.19) because while a load F increases it bears against a part more tightly thus adding load-carrying capability. This effect would bring about a significant increase in the bearing capacity of frictional joints or shear strength of the "fibre–matrix" interface under mechanical or thermo-mechanical load.

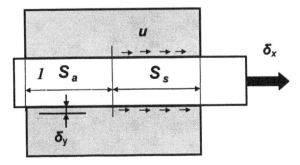

Fig. 3.19. Scheme of self-locking effect in frictional joint with auxetic deformable element

The expediency of auxetic PTFE usage in the large blood vessels prosthetics is shown (Caddock and Evans 1995). Exploitation of porous auxetics as filters and sizing screens attracts with the adjustable filter dimensional size selectivity. The pore size increases when load increases thus allowing thereafter the larger particles pass through.

It is possible to regulate the acoustic properties of the products using auxetics (Howell et al 1994, Alderson et al 1997, Lim et al 2014). The sound absorption coefficient and the loss modulus in porous materials where $v < 0$ are higher in comparison with conventional porous materials of the same porosity and density. This is due to the fact that the acoustic properties of the material are determined by the propagation velocity ratio of the longitudinal and transverse waves, depending on the Poisson ratio: $v_l / v_t = \sqrt{(1-2v) / 2(1-v)}$. If for the conventional isotropic materials the v_l / v_t ratio does not exceed $1/\sqrt{2}$, then it reaches $\sqrt{3}/2$ in auxetics. The British company Auxetix Ltd produces a protective cloth – auxetic Zetix which is not torn under the effect of the shock wave.

The absorption coefficient of ultrasound is 47 dB/cm for the auxetic UHMPE and is 1.5 times higher than in the conventional polyethylene foam (Alderson et al 1997). The auxetic crystals can be used in different sensors, amplifying a feedback of piezoelectric placed between two auxetic electrodes.

3.4.4 Compensation of thermal expansion in composites with auxetic phase

The possibility of the thermal deformations compensation with the exhibiting negative Poisson's ratio auxetics was hypothesized in the paper (Pleskachevsky and Shil'ko 2003). The mesomechanical approach is applicable for the tracing of such material structure. The idea of this method is to evaluate thermal deformations of the material using its model in the form of a of periodicity cells collection.

The finite element models covering four types of material were built: (*a*) layered (sandwich-type), (*b*) chess, (*c*) disperse-filled with disc reinforcing particles and (*d*) porous (Fig. 3.20). Coefficient of thermal expansion (CTE) of the materials was specified as $\alpha = 10^{-5}$. The thermal deformations calculation was performed assuming a plane-deformed state of the model. The displacement in the X and Y axes was used as output parameter. The fragment consisting of 5x5 cells was taken as a representative volume.

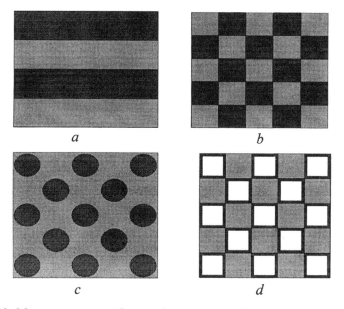

a b

c d

Fig. 3.20. Mesostructures with auxetic components having zero CTE: layered structure (*a*); chess (*b*); matrix with disk filler (*c*); porous (*d*)

The study of the layered structure thermal expansion shows that the auxetic phase having $v = -0.5$ reduces the displacement u_x 1.22 in times and u_y 1.44 in times if to compare with the phase having $v = 0.48$ (substantially incompressible material such as rubber). The most significant effect of the CTE reducing has been achieved in case of the auxetic phase having the theoretically permissible minimal value of $v = -1$ (reduction 3.62 for u_x and 5.60 for u_y correspondingly).

Similar calculations done for the homogeneous (as Young's modulus) material consisting of square shaped alternating auxetic ($v = -0.5$) and non-auxetic cells showed a decreasing of CTE in the X and Y axes equal to 1.22 and 1,28 correspondingly. It is found that by varying Poisson's ratio in a range of allowable values from 0.5 to -1.0, thermal displacement close to zero are achieved at $v = -1$.

Similar calculations for porous material formed by alternating square shaped cells, also showed a possibility of the CTE lowering. The composite materials with auxetic phase in the form of discs demonstrate a CTE reduction in the 1.34 and 1.41 times, and matrix in 1.81 and 2.45 of the times.

Thus, a comparison of these investigated models thermal expansion shows that the greatest reduction in the CTE corresponds to the homogeneous auxetic structure with a minimum value of Poisson's ratio $v = -1$. The influence of the material structure is smoothed with the Poisson's ratio increasing and becomes insignificant when $v > 0.3$.

3.4.5. Reversible ransformation of porous material into auxetics

The stress analysis has enabled us to clarify the mechanisms of adaptation to the external action, and to disclose, to a certain degree, the effect of structure on formation of the optimum feed back reaction. As was shown previously the inverted or re-entrant cell structure of porous auxetics may be obtained by isotropic permanent volumetric compression of the conventional foam, resulting in micro-buckling of the cell walls. Below we analytically predict the deformative behaviour of porous materials under uniaxial tension and compression.

For open-cell flexible cellular materials, Poisson's ratio can be determined by a rod type structural unit with chaotically oriented cubic cells, as presented in Fig. 3.21. It is worth mentioning that such a kind of unit cell model has been simulated in reference (Gibson and Ashby 1982). However, a cubic, not a spherical, structural unit had been used. Also, shear deformation of the rods had not been taken into account.

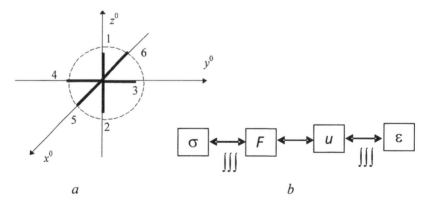

a   *b*

Fig. 3.21. Structural unit (*a*) and simulation procedure (*b*) of flexible cellular plastics: σ, ε – stress and strain tensor components; F – force acting on the rod end; u – displacement of force application point relatively to the rods; $\int\!\!\int\!\!\int$ – averaged over direction

The rods of this structural unit are directed normally to the cubic planes. Symmetry of the element allows one to represent the displacement of the force application points (ends of rods) relatively to the rod joints through the deformation tensor components:

$$\Delta x_{L1} = x_{L1} - L = L\varepsilon_{z^0 z^0}, \quad y_{L1} = \frac{L}{2}\sqrt{\gamma_{x^0 z^0}^2 + \gamma_{y^0 z^0}^2},$$

$$\Delta x_{L3} = x_{L3} - L = L\varepsilon_{y^0 y^0}, \quad y_{L3} = \frac{L}{2}\sqrt{\gamma_{x^0 y^0}^2 + \gamma_{y^0 z^0}^2}, \qquad (3.12)$$

$$\Delta x_{L5} = x_{L5} - L = L\varepsilon_{x^0 x^0}, \quad y_{L5} = \frac{L}{2}\sqrt{\gamma_{x^0 y^0}^2 + \gamma_{x^0 z^0}^2},$$

where L is the structural unit rod length; x_{Li}, y_{Li} are the coordinates for the end of the *i*-th rod (*i* = 1...6) in the *xy* coordinate system; the *x* axis is directed longitudinally to the *i*-th rod in the non-deformed state (Fig. 3.22).

Equations (3.12) refer to deformations for which the Cauchy relations are satisfied. Here, the parameter L can be related to the solid state volumetric fraction by the following equation:

$$V_f = \frac{V_m}{V} = \frac{9}{2\pi q^2}, \qquad (3.13)$$

where V_m is the volume of rods in the structural unit; V is the structural unit total volume before deformation; q is the rod length L to its cross sectional side length r ratio. For simplification we neglected the volume of the nodes (rod joints) and assumed that the rods have a square cross-section.

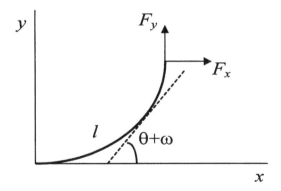

Fig. 3.22. Scheme of cantilever beam under large bending

During further calculations we have estimated that the simulation results do not depend on the r value. So we may assume that $r = 1$, $L = q$.

Let us assume that in the coordinate system XYZ uniaxial strain is defined as $\varepsilon_{nm} = f(t)$ (other components of strain are equal to zero). The system XYZ position relative to the system $x^0 y^0 z^0$ is defined by Euler's angles $\beta_1, \beta_2, \beta_3$. Once the function $f(t)$ and Euler's angles are known, these define the deformation components in the $x^0 y^0 z^0$ system (Fig. 2). Then, the displacements (3.12) can be written as follows:

$$y_{Li} = \varepsilon_{nm}(t)\eta_i(\beta_1, \beta_2 \beta_3), \quad \Delta x_{Li} = \varepsilon_{nm}(t)\xi_i(\beta_1, \beta_2 \beta_3). \tag{3.14}$$

Here $\eta_i(\beta_1, \beta_2, \beta_3)$, $\xi_i(\beta_1, \beta_2, \beta_3)$ are the Euler's angle functions which are related by recalculating the tensor components under coordinate axis rotation. For the determination of forces $\bar{F}_i$ acting at the ends of the rods by the set deflections, it is necessary to solve a large flexure problem of a cantilever beam taking into account material viscosity. At the same time, to describe deformation of the low-density porous materials ($V_f < 0.1$) it can be assumed that the rod is deformed equally over all length L. The viscoelastic behaviour of the rod material is described by Rzhanitsyn's relaxation function

$$R(t) = Ae^{-\beta t}t^{\alpha - 1}, \tag{3.15}$$

where t is time; and A, α, β are the kernel parameters.

The stress/strain relations are determined by the following equation

$$s_{px} = 2G_f\left(v_{px} - \int_0^t R(t - \tau)v_{px}(\tau)d\tau \right), \quad \sigma = 3K_f\varepsilon. \tag{3.16}$$

where $s_{px}, v_{px}, \sigma, \varepsilon$ are the deviatoric and spherical parts of the stress and strain tensors; G_f, K_f are the shear and bulk moduli of the material.

For the beam deformations, let us assume that

$$\varepsilon_{ll} = \varepsilon_0(l) + \lambda\theta'(l), \quad \varepsilon_{\lambda l} = \frac{1}{2}\omega(l), \tag{3.17}$$

where l is the coordinate referred along the rod median in the deformed state; λ is the coordinate referred perpendicularly to l; θ is the rotation of the rod cross-section connected with flexural strain; $\theta' - \theta$ is the derivative of the l coordinate; ω is the rod cross-section turning angle as a function of shear strain; ε_0 is the deformation of the centre line passing through the centre of gravity under tension or compression.

The allowance for flexural, shear and tensile-compression strains helps to describe deformation of 'short' rods when their length is commensurable with the cross-sectional side length. For an arbitrary cross-sectional shape, the following expressions are valid

$$M = \iint_S \sigma_{ll}\lambda dS, \quad P = \iint_S \sigma_{ll} dS, \quad Q = \iint_S \sigma_{\lambda l} dS, \tag{3.18}$$

where M is the bending moment; Q, P are the transverse and longitudinal forces. Therefore, the equilibrium equations for the cantilever rod for the large deflection case will take the form

$$Q = F_y\cos(\theta + \omega) - F_x\sin(\theta + \omega), \quad P = F_x\cos(\theta + \omega) + F_y\sin(\theta + \omega),$$
$$M = F_y(x_L - x) - F_x(y_L - y). \tag{3.19}$$

Substituting equations (3.16) and (3.17) into (3.18) gives:

$$\omega = \frac{-k}{G_fS}(F_x\sin(\theta + \omega) - F_y\cos(\theta + \omega)) + \int_0^t R(t - \tau)\omega(\tau)d\tau,$$

$$\varepsilon_0 = \frac{1}{E_fS}(F_y\sin(\theta + \omega) + F_x\cos(\theta + \omega)) + \int_0^t R(t - \tau)\varepsilon_0(\tau)d\tau, \tag{3.20}$$

$$\theta' = \frac{1}{E_fJ}\left(F_y(L + \varepsilon_{nm}\xi - x) - F_x(\varepsilon_{nm}\eta - y)\right) + \int_0^t R(t - \tau)\theta'(\tau)d\tau,$$

$$x' = \cos(\theta + \omega), \quad y' = \sin(\theta + \omega).$$

Here J, S are the second moments of the area and the cross-sectional area of the rod, correspondingly; E_f is Young's modulus of rod material; k is the coefficient complying with non-uniformity of tangential stress distribution over the cross-sectional area. In our calculations we assumed $k = 1$.

Therefore, a system of equations was obtained for the five unknown coordinates l and time functions. Let us apply the following boundary conditions: $\theta(0, t) = x(0, t) = y(0, t) = 0$. In (3.20) η, ξ are constants. Solution of these combined equations using the finite difference method allows us to obtain the coordinates of the free end of rod as a function of five variables, viz:

$$x_L = x(L,t) = f_x(F_x, F_y, \eta, \xi, t), \quad y_L = y(L,t) = f_y(F_x, F_y, \eta, \xi, t). \quad (3.21)$$

During computation of (3.20) it was taken into account that the l coordinate differentiation is made in the deformed state. Therefore, the increment of the l parameter was assumed equal to $dl = (1 + \varepsilon_0) \dfrac{L}{n_0}$, , where n_0 is a discretization number.

The solution of (3.20) was carried out for a specified t. It should be mentioned that the structure of Rzhanitsyn's relaxation function (3.15) causes the integral terms in (3.20) to contain θ, γ and ε_0 functions which were defined during the previous steps. The conditions for calculation of the required forces are of the type

$$\begin{cases} f_x(F_x, F_y, \eta, \xi, t) = L + \varepsilon_{nm}(t)\xi, \\ f_y(F_x, F_y, \eta, \xi, t) = \varepsilon_{nm}(t)\eta \end{cases} \quad (3.22)$$

The solution of equations (3.20) and (3.22) was obtained numerically with the help of MathCad® 7.0 software. The system of nonlinear equations was solved using Newton's method. As the initial approximation we took the solution of the previous step. Therefore, we obtain the functions $F_x(\eta, \xi, t)$, $F_y(\eta, \xi, t)$ which can be presented as follows:

$$F_x = C_{x1}\xi + C_{x2}\eta + C_{x3}\xi\eta + C_{x4}\xi^2 + C_{x5}\eta^2 +$$
$$+ C_{x6}\xi^2\eta + C_{x7}\xi\eta^2 + C_{x8}\xi^3 + C_{x9}\eta^3 + C_{x10}\xi^2\eta^2. \quad (3.23)$$

At a given t, the coefficients C_{xj}, C_{yj} ($j = 1 \ldots 10$) can be defined by standard regression procedures. The stress tensor components are related through the forces (3.23) as follows:

$$\sigma_{x^0x^0} = F_{x1} \frac{1}{\pi(L + \varepsilon_{nm}\xi_2)(L + \varepsilon_{nm}\xi_3)}, \quad \sigma_{y^0y^0} = F_{x2} \frac{1}{\pi(L + \varepsilon_{nm}\xi_1)(L + \varepsilon_{nm}\xi_3)},$$

$$(3.24)$$

$$\sigma_{z^0z^0} = F_{x3} \frac{1}{\pi(L + \varepsilon_{nm}\xi_1)(L + \varepsilon_{nm}\xi_2)},$$

$$\sigma_{x^0y^0} = F_{y1} \frac{1}{\pi(L + \varepsilon_{nm}\xi_2)(L + \varepsilon_{nm}\xi_3)} \frac{\varepsilon_{x^0y^0}}{\left(\varepsilon_{x^0y^0}^2 + \varepsilon_{x^0z^0}^2\right)^{1/2}}, \quad (3.25)$$

$$\sigma_{x^0z^0} = F_{y1} \frac{1}{\pi(L + \varepsilon_{nm}\xi_2)(L + \varepsilon_{nm}\xi_3)} \frac{\varepsilon_{x^0z^0}}{\left(\varepsilon_{x^0y^0}^2 + \varepsilon_{x^0z^0}^2\right)^{1/2}},$$

$$\sigma_{y^0z^0} = F_{y2} \frac{1}{\pi(L + \varepsilon_{nm}\xi_1)(L + \varepsilon_{nm}\xi_3)} \frac{\varepsilon_{y^0z^0}}{\left(\varepsilon_{x^0y^0}^2 + \varepsilon_{y^0z^0}^2\right)^{1/2}} \cdot$$

The stresses for the *XYZ* system were then redefined. Because of the chaotic orientation of the unit cells, the stress tensor components should be averaged over direction (Euler's angles)

$$\sigma_{nm} = \int\limits_0^\pi \int\limits_0^{2\pi} \int\limits_0^{2\pi} \sigma_{nm}(\beta_1, \beta_2, \beta_3) \frac{\sin\beta_3}{8\pi^2} d\beta_1 d\beta_2 d\beta_3. \tag{3.26}$$

Therefore, for the known stress to time dependence, we defined time dependencies of the stresses in a representative volume of the material.

As an example of using the above technique, let us examine the stress-strain state of an elastic porous material based on high density polyethylene (HDPE). Experimental data for HDPE were obtained from (Goldman 1979): $G_f = 237$ MPa; $K_f = 1402$ MPa; $A = 0.022$ s^{-1}; $\beta = 2.995 \times 10^{-5}$ s^{-1}; $\alpha = 0.175$.

Averaging in all possible loading directions (3.22) makes the simulated material isotropic at the macroscopic level. The $\tau(\gamma)$ function therefore characterizes the dependence of stress on strain deviator components $\tau(\gamma) = s_{nm}(2\upsilon_{nm})$. Thus, if functions $\tau(\gamma)$ and $p(\Theta)$ are known, it is possible to simulate isotropic material behaviour at an arbitrary homogeneous stress-strain state. Hence, for a uniaxial stress ($\sigma_{zz} \neq 0$) the following relations are true

$$2\upsilon_{ZZ} = \frac{4}{3}\varepsilon_{ZZ}(1+\mu); \quad \Theta = (1+\varepsilon_{ZZ})(1-\varepsilon_{ZZ}\mu)^2 - 1;$$

$$s_{ZZ} = \frac{2}{3}\sigma_{ZZ}; \quad \sigma = \frac{1}{3}\sigma_{ZZ}. \tag{3.27}$$

We introduce the transverse strain factor $\mu = -\dfrac{\varepsilon_{xx}}{\varepsilon_{zz}}$ which is analogous to Poisson's ratio in the linear elasticity region. Making allowance for a large bending flexure of the ribs where μ depends on strain ε_{ZZ}, this dependence is determined by the following equation:

$$\tau\left(\frac{4}{3}\varepsilon_{ZZ}(1+\mu)\right) = 2p\left((1+\varepsilon_{ZZ})(1-\varepsilon_{ZZ}\mu)^2 - 1\right). \tag{3.28}$$

Under stretching, μ also decreases rapidly when the strain reaches ε_{cr}. In addition, the $\mu(\varepsilon_{ZZ})$ dependence rapidly passes on the horizontal plateau $\mu(\varepsilon_{zz}) = const = \nu^0$, where ν^0 is defined as

$$\nu^0 = \frac{3K - 2G^0}{6K + 2G^0}, \tag{3.29}$$

where K is the foam bulk modulus defined by the initial part of the $p(\Theta)$ curve; G^0 is the shear modulus defined by the $\tau(\gamma)$ curve.

It was found that the $\mu(\varepsilon_{ZZ})$ function does not depend on the strain rate. The dependence of the factor μ on the longitudinal strain ε_{ZZ} at stretching (*a*) and compression (*b*) of an elastic cellular plastic based on HDPE ($V_f = 0.01$) is presented in Fig. 3.23. Under compression the strain reaches some critical value ε_{cr} and μ rapidly decreases and becomes negative at $\varepsilon_{ZZ} > 0.9$ %. Such an anomaly of the elastic behaviour was observed experimentally in polymer foams (Lakes 1987). Our investigations showed that this effect may occur in cellular materials with a tetragonal cell form when the cell ribs buckle inward or in a honeycomb microstructure.

At small strains, μ remains constant and coincides with Poisson's ratio. Under compression, the strain reaches some critical value ε_{cr} and μ rapidly decreases and becomes negative at $\varepsilon_{ZZ} > 0.9$ %. Under stretching, μ decreases rapidly when the strain reaches ε_{cr} and $\mu(\varepsilon_{ZZ})$ dependence passes on the horizontal plateau. By defining the $\mu(\varepsilon_{ZZ})$ function, the dependence of stress σ_{ZZ} on strain ε_{ZZ} can be obtained as

$$\sigma_{ZZ}(\varepsilon_{ZZ}) = \frac{3}{2}\tau\left(\frac{4}{3}\varepsilon_{ZZ}\left[1 + \mu(\varepsilon_{ZZ})\right]\right). \tag{3.30}$$

At small strains, stability of μ allows us to determine the correlation between ε_{cr}, Θ_{cr} and γ_{cr}:

$$\varepsilon_{cr} = \frac{1}{1-2v}\Theta_{cr}, \qquad \gamma_{cr} = \frac{4(1+v)}{3(1-2v)}\Theta_{cr}. \tag{3.31}$$

To examine the applicability of the theoretical model for foam deformation properties, we compared the calculated and experimental values of the relative Young's modulus E/E_f and critical strains ε_{cr} proceeding from the following considerations: the majority of experimental data on deformation of elastic foams are based on their uniaxial compression behaviour; the calculated stress/strain dependence and the experimental behaviour are almost linear at $\varepsilon < \varepsilon_{cr}$. As it was shown in (Hilyard and Cunningham 1987), $\sigma_{ZZ}(\varepsilon_{ZZ})$ dependence at $\varepsilon_{ZZ} > \varepsilon_{cr}$ to a certain degree is conditioned by inhomogeneity of the inner structure of the material.

During definition of Young's modulus E of an elastic cellular plastic, we considered that the rod cross-section turning angles are small ($\cos(\theta + \omega) \approx 1$, $\sin(\theta + \omega) \approx 0$) and we do not consider rod viscosity. In this case, the solutions can be obtained in the analytical form. For the relative Young's modulus, we have

$$\frac{E}{E_f} = V_f\frac{36 + V_f\pi(7 + 4v_f)}{216 + 3V_f\pi(9 + 8v_f)}, \tag{3.32}$$

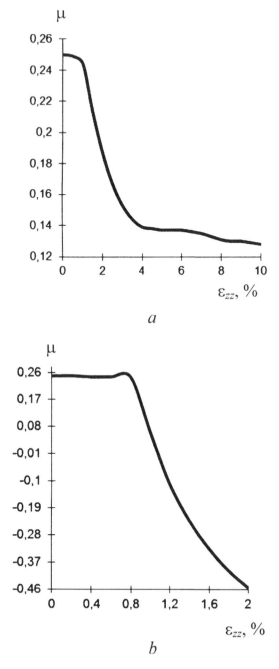

Fig. 3.23. Dependence of transverse strain factor μ on longitudinal strain ε_{ZZ} under tension (*a*) and compression (*b*) of flexible cellular plastics

where v_f is the solid phase Poisson's ratio. In particular, for the elastic polymer material we assume that $v_f = 0.49$. Equations (3.18) and (3.23) yield an approximate expression for the critical strain ε_{cr}

$$\varepsilon_{cr} = \frac{V_f \pi^3 \left[72 + V_f \pi \left(9 + 8v_f \right) \right]}{72 \left[36 + V_f \pi \left(7 + 4v_f \right) \right]}.$$ (3.33)

The dependence of the relative Young's modulus E/E_f on the relative solid volume fraction V_f for the elastic foam is presented in Fig. 3.24.

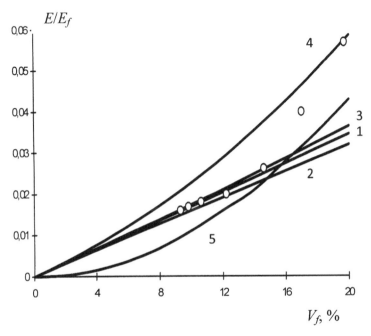

Fig. 3.24. The dependence of relative Young's modulus E/E_f on the solid phase volumetric fraction V_f

In Fig. 3.24, curve *1* corresponds to equation (3.31). Curve *2* agrees with the results obtained in (Warren and Kraynik 1987). Curve *3* meets the results obtained in (Beverte and Kregers 1987) using the semi-axes hypothesis. Curve *4* corresponds to the analytical expression

$$\frac{E}{E_f} = \frac{V_f}{3} \left(1 - 2v' \right) = 0,16V_f,$$ (3.34)

obtained in (Gibson and Ashby 1982).

Here v' is the Poisson's ratio of the material dependent on the number of rods in structural unit N. For simulation of mechanical behaviour of the rubber foam (Gibson and Ashby 1982), we used a structural element with $4 < N < 8$, when $v' = 0.26$. Curve 5 corresponds to the empirical relation for the relative Young's modulus of foam rubbers (Hilyard and Cunningham 1987)

$$\frac{E}{E_f} = \frac{V_f}{12}\left(2 + 7V_f + 3V_f^2\right). \tag{3.35}$$

The circles in Fig. 3.24 reflect experimental data for the foam rubber (Lederman 1971). This comparison proves that the proposed technique makes it possible to predict quite accurately elastic properties of the material at $V_f < 0{,}15$.

3.4.6. Self-assembling of auxetic structure in porous materials

There is an interest in compression-driven self-assembly as a means to create auxetic porous structures at the nanoscale. Mesomechanical (in the scale of the separate cells) description of cellular structure is time-consuming but a very informative method. A possibility for determination of Poisson's ratio during special thermomechanical treatment of basic porous material when convex structural cells transform into concave ones as shown in Fig. 3.25, is an advantage of the mesoscopic model.

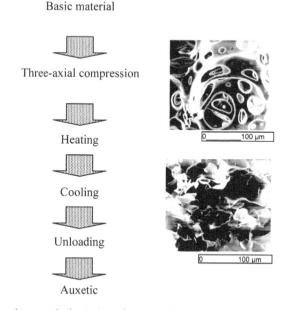

Fig. 3.25. A scheme of obtaining the auxetic material using transformation of their basic structure into the inverted one with concave cells. Microscopy of a porous polyurethane fragment with magnification 50*

Previously, the determination of v as a function of the transverse and longitudinal strain was achieved for the case of compression of the sample made of a one-phase material with known values of Poisson's ratio (Fig. 3.26a). The geometrical sizes of the rectangular sample are L_x = 50 μm, L_y = 250 μm; the compressive strain is ε_y = 0.5 %. The calculated results are shown in Table 3.7. It should be noted that the technique has an acceptable accuracy which increases as the friction between the sample and the plate decreases. This fact is explained by a free slip of the contact surfaces.

Table 3.7. The calculation results of the transverse displacements.

The number of node	u_x, μm			
	$f = 0.1$		$f = 0.5$	
	Left side	Right side	Left side	Right side
1	-0.0507	0.0498	-0.0510	0.0493
2	-0.0507	0.0498	-0.0510	0.0493
3	-0.0507	0.0498	-0.0510	0.0494
4	-0.0507	0.0498	-0.0509	0.0494
5	-0.0506	0.0499	-0.0508	0.0495
6	-0.0506	0.0499	-0.0508	0.0496
7	-0.0506	0.0500	-0.0507	0.0498
8	-0.0506	0.0503	-0.0508	0.0503
9	-0.0504	0.0483	-0.0504	0.0499
10	-0.0373	0.0370	-0.0471	0.0405
u_x	-0.04929	0.04846	-0.05045	0.0487
u_x, average	0.048875		0.049575	

For calculation of the effective elastic characteristics of the porous material mesofragments we replace the real structure by a system of cells of regular polyhedrons. The transformation of the porous material into the auxetic one appears to be possible at bulk compression V_{in}/V_{tr} equal to 1.4÷4.8 where V_{in}, V_{tr} are the volume of the initial and inverted structure respectively. The best results are achieved at V_{in}/V_{tr} = 3.3÷3.7. This agrees with the data derived for the foamed polyurethane and copper sponge.

The simulation allows us to describe cell transformation at the expense of free volume due to connection of structural units providing the required deformation mode.

According to the mesomechanical approach, some systems of regular polyhedrons presented in Fig. 3.25, were constructed for calculation of Poisson's ratio v during structural transformation under compression (Fig. 3.26). In the numerical example, we give the following initial data for the solid phase of the porous material: $E = 1$ GPa, $v = 0.1$; the sizes of the fragment 240×280 μm and the periodic cell 34×34 μm, the friction coefficient on the contacting surface with the rigid plates is $f = 0.5$.

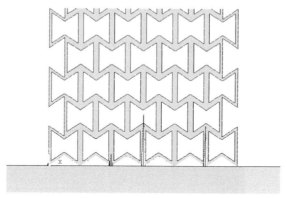

Fig. 3.26. Distribution of contact pressure under deformation of the porous structure (vertical displacement $u_y = 0.1$ μm)

Besides the linear elastic solid phase, we have assumed a physically nonlinear multimodular solid phase. In the last case, the stepwise dependence of Young's modulus on the stress component has been used (3.11).

We then simulated deformation of the initial structure with rectangular cells to analyze the formation of auxetic properties under compression of traditional porous material. To increase the accuracy, Poisson's ratio was determined by averaging the displacements for the left and right sides of the model structure fragment. According to Table 3.8, the results in the case of a multimodule solid phase seems to be more stable than for $E = $ const and at less expressed auxetic properties (stability loss of the porous fragment made of multimodule material is absent at compression displacement $u_y = 14.0$ μm).

Table 3.8. The calculation results of Poisson's ratio v.

u_y, μm	1.4	2.8	7.0	14.0*	21.0	28.0	35.0	42.0
$E = $ const	-0.040	-0.054	-0.085	-0.49	-0.180	-0.222	-0.130	-0.291
$E = E(\sigma)$	-0.0146	-0.0195	-0.0340	-0.076	-0.080	-0.100	-0.118	-

*stability loss of porous material fragment.

The dependences in Fig. 3.27 were shown in a dimensionless form (compression level was taken as a ratio of normal displacements to the height of the porous material fragment u_y/b) for comparison under different conditions of loading. It can be seen that instability of solution is observed at a step-by-step loading of the porous material with hexagonal cells at a deformation level of 5 %. The porous material with hexagonal cells and multimodular solid phase coincide closely. The solution is not converged for the porous material with square cells at deformation level more than 15 % with local unstability in the range 2.5-7.5 % and for multi-module solid phase at deformation level more than 10 %.

The solution is not converged at the deformation level more than 5 % for the concave cells with linear elastic and multimodular solid phase. According to the previous stress history the solution is not converged at the deformation levels greater than 7.5 % and 3 % for the square and concave cells respectively.

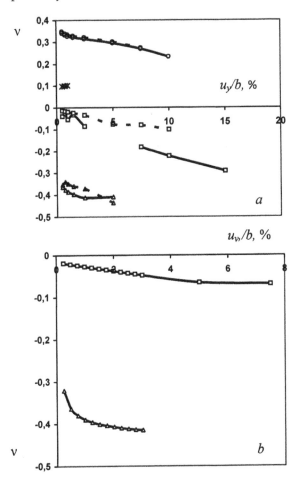

Fig. 3.27. Dependence of Poisson's ratio on compression level: *a* – step-by-step loading; *b* – accounting the previous stress history for the porous material with square cells (squares), hexagonal cells (generated angle α = 60°) (triangles), circular cells (stars) and with multimodular solid phase (hatches)

The analysis of the stress-strain state of these cellular structures for various deformation levels shows that Poisson's ratio is near to zero at the initial stress state but decreases significantly under compression of the material, which its solid phase has a constant elasticity modulus. The predicted auxetic behaviour is due to generation of the concave cells at the determinative compression level. Poisson's ratio decreases for the structure with the given concave cells transferring into a plateau. At significant deformation, the solution is not converged due to closing of the cell edges.

At the macroscale the model of the cell structure is unstable. This may result in a displacement of the fragment (in the given example this takes place at compression level u_y = 14 µm). For obtaining a stable solution, it is necessary to take into consideration the previous stress history of the contact friction process (Fig 3.27,*b*). The account of the previous stress history is also important for calculating the auxetic self-locking mode at the conditions of contact compression and shear (Shilko et al. 2008a).

Self-assembling high-strength and rigid materials of small density are of great interest like Langmuir films. This may be reached by the auxetic porous material 'construction' on the micro- and nano-size level. It is important that the value of adhesion forces F increases essentially at decreasing of the gap H between solid surfaces. The values of the adhesion force are shown in Table 3.9 for two pairs of polymers and three values of the gap H according to

$$F = \frac{A_{12}}{6\pi H^3},$$
(3.36)

where A_{12} is the Hamaker constant and H is the distance between surfaces.

Table 3.9 Estimation of adhesion stress F between surfaces of polymers (Ainbinder and Loginova 1976)

Material	F, MPa			A_{12}, Erg
	$H = 10A$	$H = 5A$	$H = 4A$	
Polytetrafluorethylene – Polyimide	7.38	51.1	115.3	$1.39 \cdot 10^{-12}$
Polycaproamide – Polycaproamide	7.30	58.3	113.9	$1.37 \cdot 10^{-12}$

It is seen that a sharp increase of the adhesion force takes place in nano-sized cells of the porous material. The calculations of the deformed state of the porous material subject to adhesion forces and multimodule effect simultaneously, show a possibility of self-assembling of a spontaneous, energetically preferable auxetic nano-sized structure as shown in Fig. 3.28.

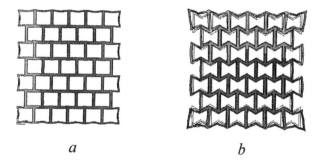

a b

Fig. 3.28. Deformation modes of the porous material with initially rectangular (*a*) and concave (*b*) shape of the cells under the action of adhesion forces

So, analytical and numerical modelling describes the cellular solid transformation resulting in microbuckling of the cell walls under certain loading conditions and providing the auxetic deformation mode. Geometrically simple mesomechanical models of the porous material based on cubic, rectangular and concave structural units may be investigated taking into account such important factors as large strains, history of loading, physical nonlinearities of solid phase, adhesive interaction as well as 'smart' properties (particularly, shape memory in auxetics (Rossiter et al 2014)).

References

Ainbinder, S.B. and A.Ja. Loginova. 1976. Adhesive interaction during friction of metal – polymer and polymer – polymer pairs. Mech. Composite Mater. 12: 734–739.

Akasaka, T. 1989. Elastic composites. pp. 315-363. *In*: T.-W. Chou and F.R. Ko (eds.). *Textile Structural Composites*. Elsevier, Amsterdam-Oxford-New York-Tokyo.

Alderson, A. and K.E. Evans. 2001. Rotation and dilation deformation mechanisms for auxetic behaviour in the a-cristobalite tetrahedral framework structure. Phys. Chem. Minerals 28: 711–718.

Alderson, K.L. and K.E. Evans. 1992. The fabrication of microporous polyethylene having a negative Poisson's ratio. Polymer 33: 4435-4438.

Alderson, K.L., A.P. Pickles, P.J. Neale and K.E. Evans. 1994. Auxetic polyethylene: the effect of a negative Poisson ratio on hardness. Acta Metall. Mater. 42: 2261–2266.

Alderson, K.L., R.S. Webber, U.F. Mohammed, E. Murphy and K.E. Evans. 1997. An experimental study of ultrasonic attenuation in microporous polyethylene. Appl. Acoustics 50: 23–33.

Anokhina, N.Yu., S.A. Bochkareva, B.A. Lyukshin, P.A. Lyukshin and S.V. Panin. 2010. Estimation of the adhesive interaction of composite material phases using the stress-strain curve. Int. J. of Nanomechanics. Science and Technology. 1: 1–11.

Ashkenazi, E.K. and E.V. Ganov. 1980. *Anisotropy of Structural Materials*. Mashinostroenie, Leningrad.

Askadskii, A.A. and V.I. Kondrashchenko. 1999. *Computational Materials Science of Polymers*. Vol. 1. Atomic and Molecular Level. Nauchnyi Mir, Moscow.

Bashmakov, V.I., T.S. Chikova and M.D. Yudin. 1983. Distribution of crack size in crystalline solids. Reports of the National Academy of Sciences of Belarus 27: 326–328.

Basistov, Yu.A. and Yu.G. Yanovskii. 1996. A hierarchic adaptive model for identification of the equations of state of viscoelastic media. J. on Composite Mechanics and Design 4: 18–49.

Baughman, R.H. and D.S. Galvao. 1993. Crystalline networks with unusual predicted mechanical and thermal properties. Nature 365: 735–737.

Baughman R.H., J.M. Shacklette, A.A. Zakhidov and S. Stafstrom. 1998. Negative Poisson's ratio as a common feature of cubic metals. Nature 392: 362–365.

Bell, J.F. 1973. Experimental foundations of solid mechanics. pp. 1–811. *In*: C. Truesdell (ed). *Mechanics of Solids*. Vol. 1. Springer, New York.

Berlin, A.A., L. Rotenburg and R Baserst. 1992. Features of deformation of disordered polymeric and non-polymeric bodies. Polymer Science, Series A. Polymer Physics 34: 6–32.

Beverte, I.V. and A.F. Kregers. 1987. Stiffness of lightweight open-porosity foam plastics. Mech. of Comp. Mater. 23: 27–33.

Bolotin, V.V. 1984. *Life Prediction of Machines and Structures*. Mashinostroenie, Moscow.

Borst R., J. Carmeliet, J. Pamin and L.J. Sluys. 1994. Some future directions in computational failure mechanics. pp. 1–12. *In*: M.A. Kusters and M.A.N. Hendriks (eds.). *DIANA Computational Mechanics '94*. Kluwer Academic Publishers, Dordrecht.

Burukhin, S.B. and I.Yu. Babkin. 1995. Composite materials of new generation (review). Khimiya Vysokikh Energii 29: 126–132.

Bushe N. and I.S. Gershman. 2006. Compatibility of tribosystems. pp. 59–80. *In*: G.S. Fox-Rabinovich and G.E. Totten (eds.). *Self-Organization During Friction. Advanced Surface-Engineered Materials and Systems Design*. CRC Taylor and Francis, Boca Raton.

Caddock, B.D. and K.E. Evans. 1995. Negative Poisson ratios and strain-dependent mechanical properties in arterial prostheses. Biomaterials 16: 1109–1115.

Casale, A. and R.S. Porter. 1978. *Polymer Stress Reactions*. Vol. 1: Introduction. Academic Press, New York, San Francisco, London.

Christensen, R.M. 1979. *Mechanics of Composite Materials*. Wiley-Interscience, New York.

Clark, S.K. 1963. The plane elastic characteristics of cord-rubber laminates. Textile Res. J. 33: 295–313.

Dementsov A. and V. Privman. 2007. Percolation modeling of conductance of self-healing composites. Physica A 385: 543–550.

Duvaut, G. and J.-L. Lions. 1972. *Inequalities in Mechanics and Physics*. Dunod, Paris.

Evans, K.E. 1991. Auxetic polymers: a new range of materials. Endeavour, New series 4: 170–174.

Evans, K.E. and B.D. Caddok. 1989. Microporous materials with negative Poisson's ratios: II. Mechanisms and interpretation. J. Phys. D: Appl. Phys. 22: 1883–1887.

Evans, K.E., M.A. Nkansah, I.J. Hutchinson and S.C. Rogers. 1991. Molecular network design. Nature. 353: 124.

Fudji, T. and M. Dzako. 1982. *Fracture mechanics of composite materials*. Mir, Moscow.

Galaev, I.Yu. 1995. 'Smart' polymers in biotechnology and medicine. Russian Chem. Reviews 64: 471–489.

Gershman, I.S. 2006. Formation of secondary structures and self-organization process of tribosystems during friction with the collection of electric current. pp. 197–230. *In*: G.S. Fox-Rabinovich and G.E. Totten (eds.). *Self-Organization During Friction. Advanced Surface-Engineered Materials and Systems Design*. CRC Taylor and Francis, Boca Raton.

Gibson, L.J. and M.F. Ashby. 1997. *Cellular Solids: Structure and Properties*. University Press, Cambridge.

Ghosh, S.K. 2009. Self-healing materials: fundamentals, design strategies, and applications. pp. 1–28 *In*: S.K. Ghosh (ed.). *Self-Healing Materials: Fundamentals, Design Strategies, and Applications*. Wiley-VCH Verlag, Weinheim.

Goldman, A.Ya. 1979. *Strength of Constructional Plastics*. Mashinostroenie, Leningrad (in Russian).

Grima, J.N. and K.E. Evans. 2000. Self-expanding molecular networks. Chem. Commun. 17: 1531–1532.

Grima, J.N., R. Jackson, A. Alderson and K.E. Evans. 2000. Do zeolites have negative Poisson's ratios? Advanced Materials 12: 1912–1918.

Hilyard N.C. and A. Cunningham. 1994. *Low Density Cellular Plastics: Physical Basis of Behaviour*. Chapman and Hall, London.

Hirotsu, S. 1991. Softening of bulk modulus and negative Poisson's ratio near the volume phase transition in polymer gels. J. of Chem. Physics 94: 3949–3957.

Howell, B., P. Prendergast and L. Hansen. 1994. Examination of acoustic behavior of negative Poisson ratio materials. Appl. Acoustics 43: 141–148.

Ivanova, V.S., A.S. Balankin, I.Zh. Bunin and A.A. Oksogoev. 1994. *Synergetics and Fractals in Materials Science*. Nauka, Moscow (in Russian).

Kolpakov, A.G. 1985. On determination of the average characteristics of elastic frames. J. of Applied Mathematics and Mechanics 99: 969–977.

Kolupaev, B.S., Yu.S. Lipatov, V.I. Nikitchuk, N.A. Bordyuk and O.M. Voloshin. 1996. Study of composite materials with negative Poisson's ratio. J. of Eng. Physics and Thermoph. 69: 542–549.

Konyok, D.A., K.W. Wojciechowski, Yu.M. Pleskachevsky and S.V. Shil'ko. 2004. Materials with negative Poisson's ratio (The review). J. on Comp. Mech. and Design 10: 35–69.

Lakes, R. 1987. Foam structure with a negative Poisson's ratio. Science 235: 1038–1040.

Lakes, R. and K.W. Wojciechowski. 2008. Negative compressibility, negative Poisson's ratio, and stability. Physica status solidi B. 245: 545–551.

Lakes, R.S. and A. Lowe. 2000. Negative Poisson's ratio foam as seat cushion material. Cell. Polym. 19: 157–167.

Lakes, R.S. and K. Elms. 1993. Indentability of conventional and negative Poisson's ratio foams. J. Comp. Mater. 27: 1193–1202.

Landau, L.D. and E.M. Lifshits. 1986. *Theory of Elasticity*. Vol. 7 (3rd ed.), Butterworth-Heinemann, Oxford.

Lazzari, B., M. Fabrizio and A. Morro. 2002. *Mathematical Models and Methods for Smart Materials*. World Sci. Publ., Ney Jersey.

Lederman, J.M. 1971. The prediction of the tensile properties of flexible foams. J. Appl. Polymer Sci. 15: 696–703.

Lim, T-C, P. Cheang and F. Scarpa. 2014. Wave motion in auxetic solids. Physica Status Solidi B – Basic Solid State Physics, 251: 388-396.

Love, A.E.H. 1944. *A Treatise on the Mathematical Theory of Elasticity*. Dover, NY.

Lyukshin, B.A., P.A. Lyukshin, S.V. Panin, S.A. Bochkareva, N.Yu. Matolygina, N.Yu. Grishaeva and Yu.S. Strukov. 2011. Computer-aided design of filled polymeric composites for tribotechnical applications. Proc. 3-rd Int. Conf. on Heterog. Mater. Mech. China. 831-838.

Makushok, E.M. 1988. *Self-organization of Deformation Processes*. Nauka i Tekhnika, Minsk (in Russian).

Martz, E.O., R.S. Lakes, V.K. Goel and J.B. Park. 2005. Design of an artificial intervertebral disc exhibiting a negative Poisson's ratio. Cellular Polymers 24: 127–138.

Mashkov, Yu.K., O.V. Kropotin, S.V. Shil'ko and Yu.M. Pleskachevsky. 2013. *Self-Organization and Structural Modification in Metal-Polymer Tribosystems*. Omsk State Technical University, Omsk (in Russian).

Miller, W., C.W. Smith, P. Dooling, A.N. Burgess and K.E. Evans. 2008. Tailored thermal expansivity in particulate composites for thermal stress management. Physica status solidi B. 245: 552–556.

Milton, G.W. 1992. Composite materials with Poisson's ratios close to –1. J. Mech. Phys. Solids 40: 1105–1137.

Moon R. 2011. Nanomaterials in the forest products industry. pp. 226-229. *In*: Yearbook in Science and Technology. McGraw Hill, Chicago.

Morozov, E.M., M.V. Zernin. 1999. *Contact Problems of Failure Mechanics*. Mashinostroenie, Moscow.

Novikov, V.V. and K.W. Wojciechowski. 1999. Negative Poisson's ratio of fractal structures. Fizika Tverdogo Tela 41: 2147–2153.

Oden, J.T., K.S. Vemaganti and N. Moës. 1999. Hierarchical modeling of heterogeneous solids. Comput. Meth. Appl. Mech. Eng. 172: 3–25.

Overaker, D.W., N.A. Lagrana and A.M. Cuitiňo. 1999. Finite element analysis of vertebral body mechanics with nonlinear microstructural model for the trabecular core. J. Biomech. Eng. 131: 542–550.

Panin, V.E. 1995. *Physical Mesomechanics and Computational Design of Materials*. Nauka, Novosibirsk.

Panin, V.E. 2010. Foundations of physical mesomechanics of structurally inhomogeneous media. pp. 501–518. *In*: V.E. Panin, Yu.V. Grinyaev and V.E. Egorushkin. Mech. of Solids 45(4). Springer, New York.

Paturi F.R. 1974. Geniale Ingenieure der Natur. Econ Verlag, Dusseldorf-Wien.

Peel, L.D. 2007. Exploration of high and negative Poisson's ratio elastomer-matrix laminates. Physica status solidi B. 244: 988–1003.

Pleskachevsky, Yu.M., S.V. Shil'ko and S.V. Stelmakh S.V. 1999. Evolution and structural levels of materials: adaptive composites. J. of Wave-Mater. Interact. 14: 49–58.

Pleskachevsky, Yu.M. and S.V. Shil'ko 2003. Auxetics: models and applications. Proc. of the National Acad. of Sci. of Belarus. Series of Phys.l-Techn. Sciences, 4: 26–36.

Ponte Castaneda, P., J.J. Telega and B. Gambin. 2004. *Nonlinear Homogenization and Its Applications to Composites, Polycrystals and Smart Materials*. NATO Science Series. II. Mathematics, Physics and Chemistry. Kluwer Academic Publishers, Dordrecht, Boston, London.

Rossiter, J.M., K. Takashima, F. Scarpa, P. Walters and T. Mukai. 2014. Shape memory polymer hexachiral auxetic structures with tunable stiffness. Smart Materials and Structures, 23: 045007.

Rothenburg, L., A.A. Berlin and R.J. Bathurst. 1991. Microstructure of isotropic materials with negative Poisson's ratio. Nature, 354: 470–472.

Schmidt, C.F., K. Svoboda, N. Lei, I.B. Petsche, L.E. Berman, C.R. Safinya and G.S. Grest. 1993. Existence of a flat phase in red cell membrane skeletons. Science 259: 952–954.

Sendeckyj, G.P. 1974. Mechanics of Composite Materials. pp. 61–102. *In*: L.J. Broutman and R.H. Krock (eds.). *Composite Materials*. Academic Press, New York, London.

Sergievich, N.V., S.V. Shil'ko and M.D. Yudin. 2000. Autolocalization of defects in adaptive composites: statistical model of process. J. on Comp. Mech. and Design 6: 504–509.

Shang, X. and R.S. Lakes. 2007. Stability of elastic material with negative stiffness and negative Poisson's ratio. Physica status solidi B. 244: 1008–1026.

Shermergor, T.D. 1977. *Theory of Elasticity of Micro-Inhomogeneous Media*. Nauka, Moscow.

Shil'ko, S. 2011. Adaptive composite materials: bionics principles, abnormal elasticity, moving interfaces. pp. 497-526. *In*: P. Těšinova (ed.). *Advances in Composite Materials – Analysis of Natural and Man-Made Materials*. InTech, Rijeka.

Shil'ko, S.V. 1995. Friction of abnormal elastic bodies. Negative Poisson's ratio. Part 1: Realization of self-locking effect. J. of Friction and Wear 15: 19–25.

Shil'ko, S.V. and Yu.M. Pleskachevsky. 2001. The mathematical simulation of free boundary evolution in frictional contact. Boundary Elem. Commun. J. 12: 18–33.

Shil'ko, S.V. and Yu.M. Pleskachevsky. 2003. Mechanics of adaptive composites and biomaterials. Materialy, Tekhnologii, Instrument 4: 5–16.

Shil'ko, S.V., E.M. Petrokovets and Yu.M. Pleskachevsky. 2008a. Peculiarities of friction in auxetic composites. Physica status solidi B. 245: P. 591–597.

Shil'ko, S.V., E.M. Petrokovets and Yu.M. Pleskachevsky. 2008b. Prediction of auxetic phenomena in nanoporomaterials. Physica status solidi B. 245: 2445–2453.

Shil'ko, S.V., V.E. Starzhinsky, E.M. Petrokovets and D.A. Chernous. 2013a. Two-level calculation method for tribojoints made of disperse-reinforced composites Part 1. J. of Friction and Wear 34: 82–86.

Shil'ko, S.V., V.E. Starzhinsky, E.M. Petrokovets and D.A. Chernous. 2013b. Two-level calculation method for tribojoints made of disperse-reinforced composites: Part 2. J. of Friction and Wear 35: 52–61.

Schmidt, C.F., K. Svoboda, N. Lei, I.B. Petsche, L.E. Berman, C.R. Safinya and G.S. Grest. 1993. Existence of a flat phase in red cell membrane skeletons. Science 259: 952–954.

Sun, L, W.M. Huang, Z. Ding, Y. Zhao, C.C. Wang, H. Purnawali and C. Tang. 2012. Stimulus-responsive shape memory materials: a review. Mater. Des. 33: 577–640.

Svetlov, I.L., A.I. Epishin, A.I. Krivko, A.I. Samoilov, I.N. Odintsev and A.P. Andreev. 1988. Anisotropy of Poisson's ratio of single crystals of nickel alloy. Proc. of the USSR Acad. of Sci. 302: 1372–1375.

Theocaris, P.S. 1994. The limits of Poisson's ratio in polycrystalline bodies. J. Mater. Sci. 29: 3527–3534.

Theocaris, P.S., G.E. Stavroulakis and P.D. Panagiotopoulus. 1997. Negative Poisson's ratios in composites with star-shaped inclusions: a numerical homogenization approach. Arch. Appl. Mech. 67: 274–286.

Toffoli, T. and N. Margolus. 1987. *Cellular Automata Machines: A New Environment for Modeling*. MIT Press, Cambridge, MA.

Wang, Y.C., J.G. Swadener J.G. and R.S. Lakes. 2007. Anomalies in stiffness and damping of a 2D discrete viscoelastic system due to negative stiffness components. Thin Solid Films. 515: 3171–3178.

Warren, W.E. and A.M. Kraynik. 1987. The effective elastic properties of low-density foams. The Winter annual meeting of the ASME, Boston: 123–145.

Wei, G. and S.F. Edvards. 1999. Effective elastic properties of composites of ellipsoids (II). Nearly disc- and needle-like inclusions. Physica A 264: 404–423.

Wei, Z.G, R. Sandstrom and S. Miyazaki. 1998. Shape-memory materials and hybrid composites for smart systems. Part I: Shape-memory materials. J. Mater. Sci. 33: 3743–3762.

White S.R., P.H. Geubelle and N.R. Sottos. 2006. Multiscale modeling and experiments for design of self-healing structural composite materials. US Air Force research report AFRL-SR-ARTR-06-0055).

Wojciechowski, K.W. 1987. Constant thermodinamic tension Monte-Carlo studies of elastic properties of a two–dimensional system of hard cyclic hexamers. Molec. Physics 61: 1247–1258.

Wojciechowski, K., A. Alderson, K.L. Alderson, B. Maruszewski and F. Scarpa. 2007. Preface. Physica status solidi B 244: 813–816.

Zubov, V.G. and M.M. Firsova. 1962. On the features of the elastic behavior of quartz in the α-β transition. Crystallogr. Rep. 7: 469–471.

Yanping, L. and Hu Hong. 2010. A review on auxetic structures and polymeric materials. Scientific Research and Essays. 10: 1052–1063.

chapter four

Smart materials and energy problem

Global problems are problems of humanity, which can be solved only on an Earth scale. P.L. Kapitsa, an outstanding scientist and public figure, was the first who formulated the four following basic interrelated global problems of the age (Kapitsa 1981): 1) a progressive increase in global population, 2) deficiency in energy resources, 3) the depletion of sources of industrial raw materials, and 4) environmental pollution. Processes that result from the global problems develop in time in accordance with geometric sequence. This leads inevitably to the phenomenon called explosion.

The current aggravation of the energy problem has noticeably changed trends in the development, production, and consumption of materials. The necessity of the economical and social assessment of material efficiency is supplemented ever more insistently by a need for using achievements in science and technology to create materials, which satisfy the changed social expedience criterion. In this final chapter, we justify the basic trends in materials science the development of which will reduce the probability of social explosion related to deficiency in natural raw materials and energy resources.

4.1. Global Energy Problem

The current state of the global energy supply system gives rise to starting search for a new complex of energy resources, which would be sufficient, first, for retaining all habitual modes of life typical of the human population, which approaches ten billion people, and, second, for producing a huge amount of material without using which the habitual human life is merely impossible.

For the mode of life typical of the USA, the current average daily energy consumption is 11 kW per person; this value is 3.5÷5.5 kW per person for Eastern Europe countries, ~ 1.0 kW per person in India and China, and the average world daily energy consumption is ~ 2 kW per person. It is reasonable to assume that, during the nearest 10÷20 years, this consumption will reach 4 kW per person, i.e. the world daily energy consumption of 40 TW (T – tera, 10^{12}) will become the energy criterion of the stable welfare of ten billion people. At present, the performance of all power facilities is 14 TW. The power of the of the biggest wind-power plant (Philadelphia, the USA) is 0.735 GW (G – giga, 10^9) and the power of the biggest solar-power plant (the US Air Force base in Nevada) is 14 MW (M – mega, 10^6). To provide a 10-TW energy increment using these plants, they should be constructed every day for ~ 20 years. This will require an enormous amount of materials, which will require additional energy for their production (Kahen and Lubomirsky 2008).

The evolution of the society can be traced not only by the types of materials used as was mentioned out in paragraph 1.1, but also in connection with the exploration of new kinds of energy (Fig. 4.1).

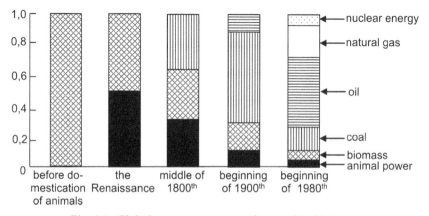

Fig. 4.1. Global energy sources in the mankind history

In the Stone Age, the combustion of wood and other biomass prevailed. Later on, people started using animal power to pull cargo. People involved wind energy in sailing, as well as in processes of grain milling and water pumping. The exploration of the energy of falling water gave rise to the production of iron in which water power dominated until a steam machine was invented. In China, water power (later on, together with coal) was the main energy source for iron melting as early as in the 5th century B.C. Nowadays, water power engineering is an important component of world power engineering; its fraction on the global scale is estimated at a few percent. The fraction of wind energy in the global energy production is substantially smaller, but this kind of energy is the most rapidly developing energy source. The scope of using nuclear energy is still comparable with the combustion of biomass partly due to the controversial stand on the reliability of nuclear reactors. The transformation of solar radiation to electric energy using photocells makes so far an insignificant contribution to the total energy resource of the globe, but holds out a promise, especially for hot-climate regions.

According to the forecasts made in (Kahen and Lubomirsky 2008), the evolution stage of solving the energy problem typical of the present will last until 2040 when science will become the driving force of the development of the society. At the current stage, utilized methods for the production, preservation, and use of energy are improved, the material and energy intensity of production reduces, the efficiency of power plants substantially increases, the amount of dissipated heat diminishes, and the consumption of solar, wind, and geothermal energy grows. In the nearest future, the following directions of power engineering will apparently become leading.

Coal purification technology will require the creation of means and corresponding materials for the temporary storage of CO_2 which will be used after payable and ecologically reasonable artificial photosynthesis methods will be developed. According to theoretical estimate, the potential of *geothermal energy* is 12 TW. The exploration of its resources, which are still untouchable, will require special methods for geothermal well drilling.

Great expectation of the humanity for a permanent source of pure and inexhaustible energy is still related to *nuclear fission*. The complexity of solving this problem was clearly demonstrated by the disasters in Chernobyl (1986) and Fukusima (2011), as well as by the abolition of the star project with the intended use of a nuclear engine by NASA in 2007. Confidence in this energy source will increase if a pure closed method for producing base nuclear fuel Th^{232} will be developed (Kahen and Lubomirsky 2008).

The consumption of *solar energy*, including wind energy and natural photosynthesis energy, progressively increases (Hirshman et al 2007). There is a need in special new-generation materials, which would play a key role in the design of solar batteries, solar energy conversion systems, and wind turbines. To produce potential 50 TW of wind energy, it is necessary to catch wind at high altitudes. This will require high-strength materials; nanotechnologies will be useful in the development of these materials. The efficiency and capabilities of artificial photosynthesis will enhance when catalysts made of cheap base metals will be created.

Since up to 10 % of electric energy will be produced from unstable sources using solar radiation, wind, surf, and flowings, the problem of its storage will aggravate. To utilize this energy, a special energy system is needed, which would possess minimum losses (e.g., the system based on superconductivity) and embrace 12 time zones. The creation of this energy system will require a huge amount of materials.

The range of fuels, which are most efficient on our planet whose atmosphere contains a substantial amount of oxygen, includes primarily oxidizable substances. The most widely used kinds of fuel are coal, oil, and gas. Oil and oil products are easy to transport and refine into products with high values of the energy density / (volume or weight) parameter.

However, oil is too expensive a raw material to combust it even partly. According to known vivid expression of D.I. Mendeleev, the author of the periodic system, to combust oil means to fire a furnace with banknotes. The non-use of the global infrastructure created to produce petrol and synthetic fuel switching are expected.

The main requirements for synthetic fuel are as follows: (1) the easiness of production from primary raw materials sources, (2) the easiness of storage and transportation, and (3) a high energy output. Table 4.1 presents the comparative data on traditional and alternative fuels obtained using this criterion.

Alcohols (ethanol and methanol), which are less toxic and more environmentally friendly than oil products, meet the above requirements to a large extent. Following these criteria, solid substances cannot serve as an alternative fuel, and among gases only those not leading to the corrosion of structural materials and easily condensed into a liquid, can be considered.

An advantage of hydrogen as a fuel is the easiness of its production by water electrolysis when a sufficient amount of Pt as the material for the electrodes is available. However, the storage of hydrogen is hampered by its corrosivity to steel. In addition, the volume density of the energy of hydrogen is fairly low; in liquid phase, it is less convenient for transportation compared to petrol and alcohols (Rand and Dell 2007).

Table 4.1 Energy output for various fuels (Kahen and Lubomirsky 2008)

Fuel	Energy output	
	mass, kJ/kg	volume, kJ/l
Coal (on average)	25.000	34.000
Wood (various kinds)	6.000–7.000	1.800–3.200
Petrol (on average)	44.000	31.000
Diesel fuel	43.000	30.000
Natural gas	50.000	32 (25.000 in liquid phase)
Methanol	19.005	15.600
Hydrogen	120.000	10 (10.000 in liquid phase)

From the viewpoint of the ease of production, safety, compatibility with structural materials, toxicity, and energy output, reduced carbon, alcohols, and synthetic petrol can be an alternative to oil products. They surpass hydrogen, as well as reduced nitrogen, boron, beryllium, aluminum, and zinc, in the above characteristics (Kahen and Lubomirsky 2008). Even the partial reduction of CO_2 to CO is sufficient for chemical industry to provide the large-scale production of the carbon alternative fuel.

The above-presented data indicate that making-up a deficiency in energy is directly related to making-up a deficiency in a variety of special materials and to the gross additional consumption of traditional materials.

4.2.   Energy Consumption for Production of Materials

The specific amount of energy consumed during the production of a material is determined by the origin of this material and the selected production method. This parameter is not an indicator of material efficiency, which depends on a more complex combination of factors, primarily, on the operating life, the possibility of recycling, the environmental friendliness of the material, the demand/production ratio, etc. Specific energy consumption characterizes to the greatest extent the role of materials in the situation caused by the energy problem. Table 4.2 presents the results of the ranking of materials following this criterion.

Concrete production is characterized by the lowest energy consumption. Taking into account that concrete is a composite based on cement and sand/gravel, cement can be considered as the most marketable material in the world. Up to two thirds of architectural and building structures are made of concrete, although only 7 % of the energy consumed by all industries accounts for concrete production. This total amount of energy is equal to about one third of the global energy production. The contribution of energy consumption to the cost of concrete is 25÷30 %, which characterizes this material as the material with the lowest energy consumption.

Table 4.2 "Energy value" of materials (Kahen and Lubomirsky 2008)

Material	Specific energy consumption for production of material, kJ/kg
Concrete	600–800
Sawn wood (veneer)	~500 (~4.000)
Glass	16.000
Steel	21.000
Steel melted from metal scrap	11.000
Aluminum (remelted)	164.000 (18.000)
Plastics, HDPE	81.000

The electrolytic melting of aluminum from ore consumes so large an amount of energy that these plants are commonly placed near to powerful hydroelectric power stations, which produce cheap electric energy in plenty. An example is Island, which is abound in river rapids, waterfalls, and cascades, as well as possesses the developed network of hydroelectric stations constructed on them. Aluminum production is the chief industry, which renders a profit to the country, although all aluminum ore is delivered to Island from far away.

The raw material base of plastics is oil. Oil-producing countries should be engaged in the stoppage of oil fuel production since the economical and social impacts of the use of oil as the raw material for plastics are many times greater than the impact of its refining to petrol and diesel fuel.

The energy problem affects the production of materials not only via restricting the amount of consumed energy, but primarily via stimulating transition to less marketable energy sources. This is confirmed by the following.

New materials become commonly in-demand after new energy sources appear or costs of using the former energy sources reduce. In transition from one energy source to another, the new source possesses, as a rule, a higher energy output than the former source. Let us compare some kinds of energy using this criterion. When one gallon (3.8 l) of petrol is combusted, 36 kW·h of energy is liberated, which is equivalent to the energy consumption for performing 500 man hours of agricultural work (about seven weeks), 50 h of the operation of a horse (one week) or the energy generated by a solar battery with an area of 60 m^2 and an efficiency 10 % during one day (6 h of solar illumination) (Kahen and Lubomirsky 2008). It can be seen that the energy output of these energy sources is less than that of oil, which is inferior only to nuclear energy in this characteristic. The energy output of a source determines the performance and structure of processes of the production of materials.

The energy problem has corrected the social expedience criterion of the production of materials. An increase in electric energy prices has caused changes in the range of materials produced due to the specification of the necessity of their use and to the energy consumption for their production processes. The recovered energy has become the criterion for the efficiency of materials. Some much advertized energy-consumed plastics and metal alloys have turned out to be noncompetitive following the price/reliability criterion.

Thus, methods for producing energy and materials are interrelated and are the chief driving force in satisfying human needs. The interdependent development of these methods will allow all population to gain certain standards and quality of life, which were unreachable or available for only a small part of humanity when the former production methods were used.

4.3. Technical-and-Economic Efficiency of Smart Materials and Technical Systems

On the face of it, energy consumption for the formation of SM and smart technical systems, which includes the creation and energy supply of feedback, should be higher than that for common materials. However, it is not always the case since, in composite SM, feedback elements can be presented by, e.g., the components of a composite, which, due to their inherent properties, play two roles, i.e., the main role and the role of a feedback element, without absorbing any additional energy. That is why the majority of SM are composite materials. It is hard to develop a method for the generalized assessment of the technical-and economic efficiency of the totality of SM without grouping the materials by the nature of feedback, purpose, conditions of operations, etc. This is a wide-ranging topic whose elaboration would inadmissibly enlarge the volume of this book. Therefore, we restrict ourselves by listing the main energy-saving advantages of SM whose efficiency can be qualitatively assessed.

The capability of SM of adapting to changes in the environmental conditions makes it possible in many cases to replace materials, which possess the reserve of the stability of properties and are thus expensive, with less stable and cheaper SM with a flexible structure, which is transformed when operating conditions change. Artificial human joints can be made of very hard high-temperature ceramics involving the polishing of the friction members in precision lapping machines, but at least the same "lifetime" of endoprostheses can be reached using much less energy-consuming polymer members, which carry a smart artificial cartilage. These endoprostheses adapt to loads applied to a joint, thus implementing lubrication mechanisms, which inherent to a natural joint.

The damage and fracture of materials always result in substantial energy and labor consumption for the repair of machines and equipment. This is a powerful stimulus for creating SM, which possess a capability of defect self-healing, indicate the critical state of an article, alleviate (within certain limits) the conditions of article operation, etc. A few examples of such SM presented in this book indicate that, at the evolution stage of solving the energy problem, no due attention has been paid to this direction of materials science. After natural reserves of enhancing the strength of materials have been practically exhausted, the problem of developing SM, which possess the property of defect self-healing, has become topical. Even its particulate solution will make it possible to reduce substantially energy consumption for the long-term operation of machines and equipment.

Units of machines and equipment with biosystems as components make it possible to implement fairly easily artificial intellect using smart responses of living cells, which are initiated by processes of metabolism. This makes the feedback structure simpler, but complicates a technical system as a whole because of a need for ensuring conditions for the metabolism of the biosystems, which are contained in it. However, with a reasonable approach to the designing of the technical facility and to the structure of biosystems, this symbiosis leads as a rule to the synergetic effect. For example, in biofilters, the clogging of the filtering element by contaminant particles is eliminated due to the vital activity of microorganisms, which consume energy from contaminants. The synergetic effect of these factors is more pronounced than the additive contributions of the processes of the mechanical trapping of the contaminant particles and of the elimination of these particles by the microorganisms to the efficiency of filtration. The energy-saving effect of the symbiosis of technical systems and biosystems is evident and will have a substantial impact on the development of many industries.

Anticorrosion SM and smart systems are convenient for development and operation since corrosion processes evolve in time and the electrodes are spatially separated when electrochemical corrosion occurs. We presented the examples of the anticorrosion protection of metal articles packed in the smart inhibited film above. Smart systems for the anode and cathode protection of oil and gas pipelines are successfully used throughout the world. Smart methods for the anticorrosion protection of concrete using bacteria, which produce calcium-containing compounds, are developed (Knoben 2011). These systems and plenty of similar systems produce the energy-saving, social, and ecological effect, which is enormous on the global scale.

Smart tribosystems represent one of the rapidly developing directions in using artificial intellect in engineering. A classical example is a friction unit in which the effect of selective transfer is implemented. This is a perfect system, which operates in a narrow load and temperature range and reduces wear to the transfer of a "third body" from one friction surface to the counterface. The possibility of the formation of a liquid lubricating film from low-melt components of antifriction materials and the availability of a wide range of self-lubricating materials widen capabilities of the development of friction units, which include feedback. The reduction of friction and wear is an inexhaustible source of energy saving the use of which will disburden the society from the incessant repair of machines and equipment.

The simplification of the design of technical articles is the most energy-saving result of the use of SM and smart technical systems. In many cases, this makes it possible to change the principle of the operation of a machine or equipment and eliminates costs due to their bulkiness. Self-lubricating materials render lubricating systems unnecessary, the self-sealing principle eliminates the system for loading contact seals, and the supply of corrosion inhibitor vapor into a closed space makes it possible to discard sealing systems, which isolate the working volume of a machine from aggressive media.

Based on this, we can determine the efficiency of the use of SM as the balance of the energies spent for their formation + repair and the energy saved during their operation:

$$E = \frac{\begin{array}{c} \text{specific energy consumption} \\ \text{for formation} \end{array} + \begin{array}{c} \text{energy consumption for repair of machines} \\ \text{due to damage and fracture of materials} \end{array}}{\begin{array}{c} \text{reduction of} \\ \text{friction and wear in} \\ \text{machines} \end{array} + \begin{array}{c} \text{reduction of} \\ \text{corrosion of} \\ \text{materials} \end{array} + \begin{array}{c} \text{self-cleaning of} \\ \text{machines from} \\ \text{contaminations} \end{array} + \begin{array}{c} \text{design} \\ \text{simplification} \end{array}}$$

For all SM and smart technical systems, $E < 1$. This value will decrease as advances in science and technology will be introduced into production. The pace of this process increases progressively. For example, despite the rise in oil prices, starting from the 1970th the fraction of natural resources per unit of the cost of industrial production decreases constantly. Therefore, the specific production yield increases even in the case of deficiency in energy, and the rise in energy resources prices serves as a stimulus for the development of alternatives.

The current level of the collective consumption of energy does not correspond to the pace of the aggravation of global problems. Social explosion can occur when the society becomes incapable of coping with the depletion of living resources. This danger can be eliminated if new kinds of energy will be discovered and the contribution of smart technologies, materials, and technical systems will sharply rise.

References

Hirshman, W.P., G. Hering and M. Schmela. 2007. Gigawatts — the measure of things to come (cell production 2006: survey). Photon Int. March: 136–166.

Kahen, D. and J. Lubomirsky. 2008. Energy, the global challenge, and materials. Materials Today 11 (12): 16–20.

Kapitsa, P.L. 1981. Scientific and Social Approach to Solving Global Problems: Experiment, Theory, and Practice. 3rd ed. Nauka, Moscow. pp. 445–466.

Knoben, W. 2011.Bacteria care for concrete. Materials Today 14 (9): 444.

Rand, D.A.J. and R.M. Dell. 2007. *Hydrogen Energy: Challenges and Prospects*. RSC Publishing, Cambridge.

Conclusion

The tendency of modern technology towards preferential use of SM and smart engineering systems represents a mechanism of energy and material saving which has no alternative methods available in the context of global issues aggravated in the 21st century. Currently, the technical reserves of strength, wear resistance, corrosion and heat resistance, as well as other characteristics determining the materials functional capabilities are practically exhausted. For this reason, SM and smart systems have been created and are being used in all areas of human activity without exception. This work has been significantly expedited after learning the computer-aided methods of FBS modeling, virtual assessment of the structure damage and deterioration of the material properties during operation, development of methods for predicting the material structural condition with moving interphase boundaries, etc.

Nevertheless, the smart systems do not play a dominant role in engineering yet, primarily, due to their relatively high cost and because society has no clear-cut understanding of the need for austerity measures with respect to saving the raw and power resources. Whenever the prices are not an insurmountable barrier, the smart systems and SM gradually become a part of the engineering domain and everyday life. The aforementioned statement is confirmed by availability of the following: smart human environment (smart house, smart clothes, smart car, etc.), smart systems for facilities management and human safety, primarily, in space, aviation, deep submergence crafts and in the marine world. Therefore, it can be said that SM and smart engineering systems are the modern paradigm of development of all (without exception) areas of material production which have been increasingly implemented at the nano- and molecular level.

The book offers the SM model which has never been applied before in the materials science. Processes of physical, chemical and biological nature initiated in the materials by external influences are the basic element of the model. The external influences are used by FBS to rehabilitate the materials structure damaged during operation. Approach to SM as cybernetic systems which are distinguished by the nature of physical and chemical mechanisms of FBS 'start-up', allowed to present (for the first time) the totality of SM in one taxonomic table 'external influence – FBS mechanism'. Its analysis shows, the capabilities of automatic compensation of the operation-induced damages of materials which have not been yet used.

The systematic analysis of the problem of smart materials developing has enabled us to trace the evolution of structural organization of artificial materials, to clarify the mechanisms of adaptation to the external action, and to disclose, to a certain degree, the effect of structure on formation of the optimum back reaction. In considered examples of composites, a description of adaptive structures is formulated as a mechanics problem on moving interfaces localizing. The study of synergetic phenomena in the nonliving nature and analogous processes in biological objects will, in our opinion, provide a possibility to find structural-and-functional prototypes of smart materials. The proposed mathematical models predict self-healing, self-assembling of porous nano-sized auxetics as well as self-reinforcing in composites and joints made of auxetic and multimodular materials.

The energy balance of the processes of SM functioning as a cybernetic system has led to the following conclusions. The current state of SM is determined by replenishment (using the FBS) of the resources of internal energy inherent in the material by nature. This occurs owing to the energy of external influences which initiate in the materials the physico-chemical processes of uncontrollable primary and secondary (under FBS control) rearrangement of the structure. Thus, the tasks of SM self-regulation can be solved by using the methods of structure 'arrangement' commonly employed in the modern materials science. The patterns of atomic and molecular interactions determined by the natural sciences are the basis for building FBS. Thus, the process of SM creation evolves from the category of heuristic flash of inspiration (as presented in the majority of publications on smart materials) into the design and engineering framework. The methodology of this creative work, energy patterns of launching physico-chemical and biological processes of reasonable rearrangement of the structure, as well as thermodynamics of non-equilibrium processes of destruction, self-assembly, technological heredity and synergism are independent subjects not covered in this book.

Time will tell, to what extent the developed approaches may prove to be fruitful and useful for the materials science and the material production as a whole.

Index